The City Street in Cameroon

A guide to street and city management

Sir Ajua Alemanji

The City Street in Camaroon

Copyright © 2021 Sir Ajua Alemanji

All rights reserved. This book may not be reproduced
in any form, in whole or in part (beyond the copying
permitted by US Copyright Law, Section 107, "fair use" in
teaching or research, Section 108, certain library copying,
or in published media by reviewers in limited excerpts),
without written permission from the author.

Published by

ISBN: Hardcover 978 0 9980857 6 0

ISBN: Paperback 978 0 9980857 7 7

Dedication

This book is dedicated to my wife, Lady Christina Nkem Amambo Ajua Alemanji. For close to five decades, she has been my wife, friend, the mother of our five children, grandmother, and mother to our extended families. She has stood by me since we got married on 23 October 1971.

When I announced to her that my creative instinct had urged me to integrate engineering with economics by returning to school, she gave me every encouragement. Both of us left Cameroon on 15 August 1977, together with our two sons, for York University in Toronto, Canada. I am deeply grateful to her for our academic success at York University where she earned a BA Honours in Economics, and where I obtained an MA in Economics.

"The human mind blossoms with infinite creative imagination"

Sir Ajua Alemanji

In March 1963, while writing an essay on space exploration for the entrance examination to the University of Nigeria, Nsukka, I dared to predict "that the United States of America and the Soviet Union [would] be forced to cooperate for the purpose of conquering space". Their joint construction of the International Space Station for the exploration of space supported my prediction. This book was written as a beacon to guide our society out of the embarrassing chaos prevailing on the streets of Cameroon, and I dare say, on the streets of other developing countries.

About the author

Chief Sir Ajua Alemanji is an accomplished engineer and economist. He was born in the village of Lebang in Bangwa in 1942 into the family of Mr. Robert Alemanji and Mama Betekong Alemnji, both blessedly remembered. The Germans named his place of birth *Fontemdorf* after Chief Fontem, and stationed their 8th army garrison there as early as 1907. Later, the British built a Native Authority Court followed by a Native Authority School.

He completed his primary education in the Native Authority School in Njenbetui before proceeding to St. Joseph's College, Sasse, Buea (1958–1963), where he obtained the West African Cambridge School Certificate. In August 1963, he enrolled in the Faculty of Civil Engineering at the University of Nigeria, Nsukka, where he studied briefly before enrolling at the prestigious Moscow Institute of Automobile and Highway Engineering.

Upon graduation in June 1970 with a Master of Science degree in Highway Engineering, he returned to Cameroon and was recruited into the Public Works Department. As assistant project manager, he designed highways, trained foremen and technical staff, and supervised reconstruction of the Kumba-Mamfe Road. He also had major inputs into the creation of the new Ekondo Titi-Mundemba Road.

In 1977, he gained admission to, and took up residence at York University in Toronto, Canada, where he studied at the Graduate School of Economics with special emphasis on transport economics. He graduated with a master's degree in Economics in November 1979.

By 1980, Sir Ajua Alemanji was back in Cameroon, having become the first Cameroonian to earn master's degrees in both Engineering and Economics. The country has benefited greatly from his expertise, knowledge and discipline. He has accomplished many government projects as Research Officer, Central Department of Contracts, Presidency of the Republic; Deputy Director, Ministry of Public Contracts and Computer Science; and as Economist and Consulting Engineer.

He served as Chairman of the Committee of Experts meeting on the evaluation of the implementation strategy of the Programme of Action of the Transport Decade for Africa (Addis Ababa, Ethiopia). He was a consultative member of the Trans-African Highway Authority. He was also the senior consulting engineer, civil works, for Inros Lackner (a German firm) for the Limbe Shipyard project.

As deputy director at the Directorate of Major Works of Cameroon (1989–1999), he organized and trained many engineers in contracting and project management. During his four decades of professional practice, he established a reputation for quality projects and cost savings. He was directly involved in the construction or rehabilitation of more than 1000 km of paved roads (Douala-Yaoundé, Bafia-Bafoussam, Bafang-Bangangte, Limbe-Idenau,

Maroua-Mokolo, Belabo-Bertoua, Bamenda-Bafoussam, etc.); construction of the Nsimalen International Airport in Yaoundé; rehabilitation of the Maroua Salack Airport; construction or rehabilitation of more than 140 km of streets in Douala, Yaoundé, Dschang, Bangangte, Kumba, Ngaoundéré, Maroua, and Limbe. All of these are examples of his dedication to nation-building. In addition, he partially supervised extension of the Dschang University Centre; construction of government ministerial buildings; extension and rehabilitation of water supply facilities in Bafoussam, Bandjoun, Beham and Bamenka; and extension of telephone exchanges in Douala and Yaoundé. For this outstanding and unparalleled record of service, he was twice decorated as Knight and then Officer of Valour of Cameroon.

Sir Ajua Alemanji has frequently published in the *Journal of the Cameroon Society of Engineers*. His articles have covered road accidents, road maintenance, and integrated approaches to development projects such as the Ekondo-Titi-Mundemba Road. His research on the Trans-African Highway (Lagos-Mombasa) served as a working document for the preparation of the NEPAD Infrastructure Action Plan. In January 2012, a special edition of *Cameroon Tribune* devoted to transport infrastructure included his research on the challenges of road construction in Cameroon.

His book *In Defense of a Tradition and the Legacy of Mafua Nkengafac Fuantem* (2006) is available at Amazon.com. His next book *The Trans-African Highway, Lagos-Mombasa: Myth or Economic Imperative?* is expected soon.

On 5 April 1997, Sir Ajua Alemanji was enthroned in the Fondom of Ndungatet as His Royal Highness Fuankeng Ajua. He is the Managing Director of AM Engineering and Contracts Consultants. He has been married for 47 years and has five children, grandchildren and a village to look after.

Contents

Dedication iii

The human mind. iv

About the author v

Abbreviations viii

Preface ix

Acknowledgements xii

Executive summary xiv

Chapter 1 – Historical background 1

Chapter 2 – The city street as a public good 5

Chapter 3 – The engineer and society 11

Chapter 4 – Traffic engineering 15

Chapter 5 – Why streets should be planned 19

Chapter 6 – Evolution of Cameroon vehicle fleet, 1980 to 2009. . 23

Chapter 7 – Demand for space in city streets 31

Chapter 8 – Use of traffic signals and signs 41

Chapter 9 – Roundabouts 45

Chapter 10 – Location of public utility services. 53

Chapter 11 – Commerce along city streets 57

Chapter 12 – Naming streets. 61

Chapter 13 – Pollution and congestion. 67

Chapter 14 – Future street design 71

Chapter 15 – Bypasses. 79

Chapter 16 – Building construction on city streets 83

Chapter 17 – Enforcement of norms and regulations. 91

Chapter 18 – Maintenance 97

Chapter 19 – The way forward 101

Conclusion. 103

Bibliography 105

Annex. 109

Abbreviations

AASHO	American Association of State Highway Officials
AASHTO	American Association of State Highway and Transport Officials
ACI	American Concrete Institute
ADT	Average Daily Traffic
ASTM	American Standard of Testing Materials
BOT	Build, Operate and Transfer
BP	British Petroleum
BSCP	British Standard Code of Practice
BS	British Standard
CBR	California Bearing Ratio
Camtel	Cameroon Telecommunications
Camwater	the Cameroon national water supply company
CFA	*Communauté Financière d'Afrique* (a unit of currency)
CP	Code of Practice
CRTV	Cameroon Radio and Television
CSE	Cameroon Society of Engineers
EMIA	*Ecole militaire interarmes* (the Military Academy)
ERP	Electronic Road Pricing
FCSE	Fellow of the Cameroon Society of Engineers
HMSO	Her Majesty's Stationery Office
ICE	Institute of Civil Engineers
MBA	Master of Business Administration
MCSE	Member of the Cameroon Society of Engineers
MAETUR	*Mission d'aménagement et d'équipement des terrains urbains et ruraux*
MINTP	Ministry of Public Works
NA	Native Authority
NASA	National Aeronautics and Space Administration
NEPAD	New Partnership for Africa's Development
NOCE	National Order of Civil Engineers
P	Paved surface
SOCAL	Standard Oil of California
SOCONY	Standard Oil of New York
SCDP	*Société Camerounaise de Depôts Pétroliers*
SOBA	Sasse Old Boys Association
TEXACO	The Texas (Oil) Company
TKC	Tonnerre Kalara Club
TR	Tractor
UCS	Unconfined Compressive Strength
UK	United Kingdom
USA	United States of America

Preface

The conceptual framework of this book is based on everyday observable events taking place on our city streets, and the desire to break the conspiracy of silence that has allowed the chaos to reign freely. The editorial in the Catholic weekly L'Effort (No. 530, 6–20 June 2012) entitled "Disorderliness is ungodly" epitomizes the state of our streets.

The City Street in Cameroon was conceived as an innovative approach to the analysis and understanding of the complex network of activities that have invaded our streets. We have adopted the philosophical view of Stephen Jay Gould in Dialogue (No. 78, April 1987, p. 13) which states that "there is a great tradition from Galileo ... that good scientists write in ways the public can understand." We do so in the full knowledge that an educated public is a valuable asset to the community.

During field research, this author came face-to-face with the incredible lack of knowledge and information on the part of the public and street users. Lane selection, especially in the roundabouts and street junctions, was observed to pose problems, especially for left-turning traffic.

Littering the city street has become an acceptable norm instead of a punishable crime, and nobody seems to feel compelled to respect even the most basic of restrictive traffic sign postings. Two glaring examples will drive home the point. A taxi driver saw the sign posting "taxi" and proceeded to park his vehicle there to shop in the neighbourhood. Upon his return, the taxi had been locked up by the council authorities for wrongful parking, and the driver was fined 25,000 CFA francs or the equivalent of US$ 50. Nobody gave him any explanation about what the sign posting meant.

Second, in the municipality of Edéa, there is a designated traffic lane for motorcycles. However, uninformed street users have transformed the motorcycle lane into curbside parking spots, thus forcing the cyclists back onto the carriageway to compete with motor vehicle traffic. The public has never been educated about why separate lanes should be reserved for motor cycles.

This book was designed and written by a T-shaped engineer as the only rational way of adjudicating the numerous socio-economic and artificial conflicts of interest prevalent on our streets. The reflections herein are based on more than four decades of practical engineering practice and 35 years of economic analysis of city management in Cameroon. They are reflections of our city life and fit squarely with the reality about what citizens experience daily on our streets.

Information, says price theory, should be gathered up to the point where the incremental cost of additional information is equal to the incremental profit that can be earned by having it. The more information the public and street users are given, the less chaos we will witness on our city streets. Conscious of the value of education, a variety of pedagogic tools have been assembled in the form of diagrams, illustrations, statistics and traffic data. This information has been included in the documentation for the following reasons:

(a) Educating the public about the real existence of the city street, its functions, uses and abuses is important;
(b) Illustrating the comprehensive and inclusive management of city councils and governments by administrators, engineers, architects, planners, and users themselves can and should be beneficial to the community;
(c) Concrete and constructive examples of meaningful planning and management policies can be extremely helpful.

Because the city street is like a conglomerate with many affiliates, various topics have been included involving engineers and community, street housing, construction concrete, planning, the national vehicle fleet, and maintenance. The narrative therefore focuses on the maze of prevailing street-related problems. These are discussed in some detail because of their impact on how the city street is currently managed and how it will be managed in the future.

An analysis of the national vehicle fleet revealed the seriousness of problems related to their exponential growth and ageing which cause congestion and pollution. The average growth rate of the fleet was 7% over 23 years of recorded positive growth. This has not been matched by provision of adequate infrastructure. This situation calls for an urgent amendment to the vehicle importation policy. That policy should limit the age of an imported vehicle to not more than 5 years from the date of manufacture.

As a pathfinder and pioneer on the subject, this study has succeeded in providing profiles about the evolution and origin of city streets that are based on African concepts and experiences. It has also relied on historical facts from Europe and America.

Street naming, the notion of the engineer and society, and the profession of traffic engineer are discussed not as footnotes but as illuminating examples of efficient management options available for our city and local governments. Their inclusion in the new and enlightened strategic policy of techno-administrative management will ensure the desired changes that will reverse the current crisis on our city streets.

In case the cynics are wondering why an entire book should be devoted to the city street, we invite them to examine the classic "without the city streets" scenario. If one morning we woke up and discovered that all the city streets had disappeared, and that in their place grass and trees had grown, the immediate consequences would be that all the utility services – electricity, water and telephones – were gone. One obvious consequence would be the absence of any form of transportation. The ensuing confusion would surpass that which gripped East Asia on the morning after the tsunami. We would thus be confined to our houses with the grim probability of starving to death.

Future desirable public policies or strategies at the local level need to provide an urban traffic system that allows the uninterrupted flow of pedestrians, cyclists, public transport vehicles, private cars, and even heavy vehicles. The organization of transport facilities must be related to urban and environmental planning and should be solely coordinated by the city or town council or government.

Efforts to educate the public on their civic duties and force them to embrace the general principle of shared space are the first steps in the right direction. As President Julius Kambarage Nyerere of Tanzania once wrote, "The most important thing for us to do now is guard our freedom to Think as well as act". This book is an attempt to think freely. As Nyerere purposely stated, "There is no standing still in life, even the most primitive biological cell experiences constant change. Society, like everything else, must either move or stagnate. ... A mind unused atrophies, a man without a mind is nothing" (cf. Freedom and Unity, 1966, p. 120).

Faced with the spiral of collapsed buildings, this study has systematically advocated that buildings will suffer spectacular damage due to surcharge loads, inadequate foundation design, lack of adequate technical supervision by qualified engineers, or causes other than natural hazards. The principles of robust design based on a strong foundation and construction must be rigorously adhered to at all times if buildings are to avoid serious structural damage or collapse.

The number of registered civil engineers in Cameroon now stands at 1446. Hence, there is no excuse for any violation of construction norms.

The choices have been put before us in this book. It is up to us to follow or ignore them. We need to bear in mind that the shortest sermon on earth is not written in the Bible but appears on the traffic signs on city streets: "KEEP RIGHT".

Acknowledgements

When confronted with the question "What has propelled you so far ahead in science and mathematics?" Sir Isaac Newton replied, "I have stood on the shoulders of giants."

If I were asked the question as to why I have written this book, I would simply say that from the day I stepped through the doors of the experimental science laboratories in Sasse College in January 1958, I began to climb the ladder of knowledge. This book was conceived and written as a testament from a T-shaped engineer to humanity in fulfilment of the role of the engineer as the problem solver in society.

There is a proverb which says "do not curse the darkness, light a candle". The chaos we have been experiencing on our city streets now threatens to drown us in darkness; this book is the candle guiding us out of this marooned situation.

In writing the book, I was inspired by the motto of the late Professor Fonlon who wrote, "not merely to recount what has been, but to share in moulding what should be". His motto is rooted in history and intuitively derived from philosophical logic: we cannot reap prosperity without sowing for posterity.

I acknowledge with profound gratitude the fact that it was Professor Bernard Fonlon (then Deputy Minister of Foreign Affairs) who selected my name from a list of meritorious students who had gained admission to universities to study engineering but who had no scholarship. He promptly offered me a government scholarship to study in the Soviet Union. He changed my life profoundly and, in so doing, had an enormous impact on the socio-economic development of Cameroon.

In writing the section on the engineer and society, I borrowed some concepts from a newspaper article offering a career guide to opportunities in engineering. Discussions on facing up to the dilemma engineers must live with were informed by my four decades of practical field and desk experience in engineering and economics.

I am deeply indebted to the authors of the Highway Engineering Handbook (edited by Kenneth B Woods et al.) for the materials on traffic engineering, channelling islands, etc. Street classification, street design, laying of underground utility service lines as well as traffic signals have their origins in Roads in Urban Areas, but the examples are of local origin.

Some of the quotations in the reflections on city street housing come from Technical Principles of Building for Safety. The remainder of that section is based on our own experience from building sites in Cameroon.

Writing a technical book of this nature, with limited access to relevant technical resources, is an arduous task. Sometimes, it is just impossible to remember every quotation that appears in the book for acknowledgement. I have therefore compiled an exhaustive bibliography to compensate for this shortcoming.

My diagrams and sketches were transformed into autoCAD for inclusion in the book by Guy Marcel Ngueuga. I thank him for the services he rendered.

Acknowledgements

The idea of the layout and design of the cover page were conceived by the author, but the graphics were perfected by Prontaprint, Wimpole Street, 95 Wimpole Street, London W1G OEJ. We are grateful for their input.

My wife, Christina Amambo Nkem Ajua Alemanji, and our five children showed me a lot of love and understanding during the writing of this book. They did not complain about my sporadic absences for field research. I thank them for their constant patience and support, without which I would never have come close to my initial target.

The priceless information captured in the photographs that illuminate the pages of this book conveys the story faithfully to readers. Photographs are from my digital camera donated for the project by our own son, Ajua Alemanji, Managing Director of AJAMSONIC. We thank him.

I am grateful and very appreciative of the patience with which Ms Nguenkam Agatha carried out the secretarial work from the initial draft to the final manuscript. I have lost count of the number of modifications I made, but she just kept on correcting them without any complaint; I owe her a debt of gratitude. I also wish to extend thanks to my daughter, Alemngoasong Akwi Ajua Alemanji, who gave up precious holiday time to help with the manuscript.

I am particularly indebted to Dr Fidelis Morfaw and his team of editors at Service Resource Africa in Prosper, Texas, USA (morfaw@gmail.com) for their outstanding professional editorial support. Their editorial prowess provided a facelift to the book.

It took more than four years to research for this book. I am indebted to the Cameroonian Ministry for Environment, Protection of Nature, and Sustainable Development for its priceless advice and support during the investigation and subsequent preparation of the manuscript. That said, I take full responsibility for the book's content and any shortcomings that may have inadvertently slipped into it.

I have always drawn inspiration from Aldous Huxley who once wrote that "the secret of genius is to carry the spirit of the child into old age, which means never losing your enthusiasm and never being afraid to try something new". We have offered something new in this book for society to learn and emulate.

Sir Ajua Alemanji
MSc Engineering, MA Economics
Consulting Engineer Emeritus

Executive summary

The public space that has evolved over time into the city street or the village street has become a pole of attraction in most communities around the world. This book was designed and written as an endeavour to educate city governments and the public on the construction and proper use of the city street. Adequate education of the citizens should enable them to make proper use of the street and reduce the environmental degradation that accompanies misuse.

Chapter 1 of the book is a historical background to the street. Its origins are traceable to early human settlements and their gradual drift from village hamlets to town communities. The account shows that the word 'street' (strasse) existed in prehistoric German, complete with equivalents in Dutch, Romanian, and Italian. Some details on how streets were built are given.

Chapter 2 discusses the street as a public good which must be used judiciously for the benefit of all. It outlines the types of street and classifies them by function into urban arterial systems, collector streets, transit routes, and local streets.

Chapter 3 is devoted to the engineer and society and defines engineer as a professional practitioner of engineering whose concern is the application of scientific knowledge, mathematics, and ingenuity to the development of solutions to technical and practical problems. The role of engineer is spelled out, along with some of the dilemmas they face.

Chapter 4 on traffic engineering discusses the problems that arise when society is unable to match demand for street and highway facilities with growth in traffic volume. Resulting congestion, accidents, and delays are shown to be the direct outcomes of the mismatch. This in turn dictates the need for legislation, regulation, enforcement, and control by city governments and public authorities.

Chapter 5 looks into why streets should be planned in the first place. Using real national examples, the author outlines the goals and objectives of street planning and discusses the consequences and impact of the failure to plan.

Vehicular traffic has a direct impact on the construction and use of the street. Chapter 6 therefore traces the evolution of the national motor vehicle fleet in Cameroon from 1980 to 2009. It grew at an average of 7% per annum during the period under review. In fact, between 1986 and 2006, the number of private cars increased by 106%. The special case of motorcycles and their propensity for road accidents is highlighted. To decongest the streets, a recommendation is made to limit the age of imported vehicles to a maximum of 5 years from the date of manufacture.

Chapter 7 is an analysis of the exponential growth in the demand for the city street, given the mix and conditions of its use, the users and their interests. The same streets must cater for long-distance trucks, delivery vehicles, shoppers, pedestrians, hawkers, and motorcycles – all in the absence of organized parking, speed limits, street signs, and properly-functioning traffic signals. Some coping strategies are proposed for adoption.

Executive Summary

As a means of giving practical effect to the coping strategies so proposed, Chapter 8 discusses the use of proper street signals and signs in the management of pedestrian and vehicle traffic on streets. Also discussed is the all-important use of average daily traffic counts for planning purposes.

Roundabouts are perhaps the most abused portions of the city street. Chapter 9 is devoted to them: their use, misuse, and abuse by both vehicles and pedestrians. The point is made that most, if not all, drivers go through driving school without learning how to use the roundabout properly. The author sketches the correct use of signs, signals, and lanes, with accompanying diagrams and pictures.

Telephone lines, electric cables, water mains, gas pipes, sewer lines, and surface water drainage must share the same street as vehicles and pedestrians. The combination can become both messy and chaotic. This is the subject of Chapter 10 that deals with the correct placement and location of public utility lines.

One of the most intractable problems of the city street is its invasion by traders who behave is if they own it. This is the subject of Chapter 11. Demand for commercial space on city streets is insatiable. This has been compounded by the invasion of motorcycles known locally as 'bendskins'. The author analyses the contribution of filling stations, garages, and billboards to the chaos on the city street.

Streets in Cameroon have virtually no names. Locations and destinations are identified by the nearest landmark or known historical figure or event (Near Moonlight Bar, *vers Carrefour des Trois Morts*, etc.). Chapter 12 deals with naming of streets and how to adopt neutral names. A long list of proposed neutral names is given in the annex to the book.

The density of both motor vehicles and the people that use them inevitably leads to pollution and congestion. This is the content of Chapter 13. It relates the congestion and pollution problem in Cameroonian city streets primarily to the indiscriminate importation of extremely old vehicles. The problem has been compounded by the use of wood as cooking fuel and the availability and use of non-biodegradable plastics. A restrictive approach to the import of motor vehicles is proposed.

Chapter 14 treats of street design and the characteristics of good streets. Whether in city centres or in the villages, proper designs should be based on actual traffic counts taken for a period of time, during a specific time of day, and during a given dry season. The data gathered are important for both construction and maintenance work.

Chapter 15 suggests a number of bypasses to relieve congestion, shorten travel time, and ease access to or through certain major cities and junctions, particularly in Bonabéri-Douala, Bamenda, and Bafoussam.

Chapter 16 deals with the painful problem of building construction on city streets where building and construction norms, guidelines, and standards are not observed. Specifications for building materials are not followed either. The specific problem of weak foundations is highlighted as this has led in part to a number of buildings collapsing. Attention is drawn to the design and construction of buildings in areas exposed to volcanic and earthquake activity.

Where a number of outdated norms and regulations exist, they are poorly enforced. Chapter 17 of the book discusses enforcement of norms and regulations. To this must be added the problem of inappropriate design and poor management of constructed structures. Real examples are cited. The organization of the practice of the profession of civil engineer in 2000 and the creation of the National Order of Civil Engineers (NOCE) holds some promise in this area.

Under maintenance of city streets (Chapter 18), the author stresses that 'everything that is conceived, designed and built has a defined lifespan contingent upon the system of maintenance adopted'. Reasons for good maintenance are advanced as well as the value of maintaining city streets. Practical guidance is given on organization of and staffing for street maintenance.

Chapter 19 sketches the way forward in 20 practical steps from origin-destination studies to the creation of an institute for the training of professional instructors of driving schools and the compulsory re-training of all existing drivers in Cameroon.

Chapter 20 is the conclusion of the book. The author sees this book as a pointer to greater exploits. Drawing on the bold vision that led the Americans and the Soviets to send humans to space, he has a futuristic vision of a bold partnership between engineers and Government where skill and authority are mutually recognized, respected, and leveraged for the benefit of all citizens.

CHAPTER 1
Historical background

The family is the fundamental unit of any community. In the African context, a family could be extended to include first, second or even third and fourth generations. In other words, a family to the African is much akin to a clan to non-Africans.

Depending on the geographical location and local habits, customs of land occupation varied from place to place. In some of the coastal areas, people tended to build their houses in clusters, whereas in the grasslands, a man's strength and position in society were determined by the number of houses in his compound and the land surface he owned or could acquire.

Each family sought to protect and shelter itself by building a home within secure boundaries. Once established, the need to feed itself became predominant, and so farming became a main occupation. Those who produced more than they could consume bartered their excess produce for what they needed. The coming together of families for the purpose of barter brought about a spider network of footpaths which led to different marketplaces for exchange.

On the one hand, although the distances between hamlets were sometimes long, within the villages, compounds were separated only by fences. As populations grew, the number of compounds which sprang up along the existing footpaths increased. On the other hand, societies whose tradition favoured the construction of unbroken chains of houses separated by a piece of land in the middle that provided access to compounds on the opposite sides may have inadvertently created the street.

As society evolved, people moved out of the villages into the towns, bringing along their settlement patterns. They sought to acquire land and to erect buildings for their families or as rental property. It was the need to commercialize goods and services in towns that gave birth to footpaths which grew into what we know today as the street. The city street is an easement or unoccupied land left or created between lines of houses or the connecting footpaths between compounds. In either case, the city street does not have a landlord or owner.

Our research has unearthed various definitions for streets, some of which date back to the fifteenth century. In German, "street" is called *strasse*; in late Latin, *strata*, or better still, *strernere*, meaning "to throw or lay stratum". A paved road was a highway with the proper names of certain ancient roads and used vaguely for a roadway or path in 1547, or a road in a town or village (comparatively wide, as opposed to a lane or alley), running between two lines of houses or shops.[1] Also, "street" may refer to the public road together with the adjacent houses, and used as a proper noun it may refer to business locations such as Wall Street or Main Street in major cities such as New York.

Etymologically, "street" refers to a road that is paved. The word appears in prehistoric German from Latin. The German *strasse* is the Dutch *straat* as well as the Italian and Romanian *strada*.[2]

Some idioms are associated with the word *street*: "The man in the street" is the ordinary man, as distinct from the expert or the man who has special opportunities of knowledge. "Not in the same street" is an informal expression meaning far inferior in terms of ability. "On the street" could refer to homelessness, but also for working as a prostitute). "Streetwise or street smart" means experienced and able to survive the ruthlessness of modern urban life, especially regarding crime. These expressions give some perspective to the origins of the word.

Early streets in Europe resembled the Roman invention of road building which made use of several layers of crushed rock or stone, each rolled hard before the next is placed. Having thus established that a street is a public road in a city, village or town, we must now turn to the uses of the city street.

---

[1] *The Shorter Oxford English Dictionary*, Third edition, Vol II, p. 2147.
[2] Street word origin, history, and etymology.

Diagram 1: Seven identifiable stages in the evolution of the street

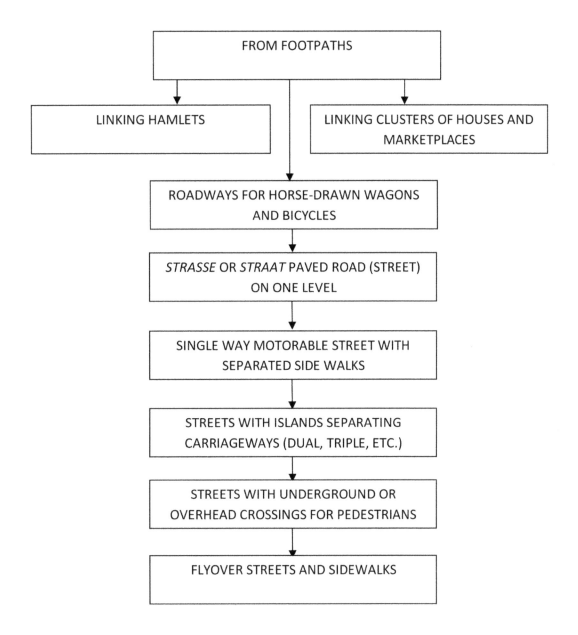

CHAPTER 2
The city street as a public good

Concerning public goods, additional consumption by one person does not imply reduced consumption by another. There are two main types of public goods. With type one, individuals can indeed vary their use of the facility (television or uncongested parks or museums). With type two goods, the individual cannot vary personal use, for example national defense or clean air.

Unfortunately, the misuse of a public good by some citizens can lead to a rationing of a limited quantity and the application of a queuing system. This is the case with taxi services where the queue ration is an all-or-nothing proposition; either the taxi operator queues and takes a passenger at the appropriate time, or the operator does not queue and therefore has no customers. Later in this text, we shall discuss the impact of queuing in taxi stands, where the present system of abusive use has resulted in total chaos.

The city street is one of those goods subject to increasing returns and used by many. There is an economic problem, however, arising from the fact that users cannot be charged as much as they are willing to pay, not because they cannot be physically excluded from the use of the good, but because any system of charging that covered costs would be inefficient. A so-called pure public good is a special case of such a good, where the marginal cost of providing an extra unit of service to an individual is zero for all units from zero to infinity. In economic terms, a city street is a public good. On the city street, as in the graveyard, all are equal.

2.1 Functions of the city street

The generic term street refers to a variety of public rights-of-way performing a variety of functions. As defined, it is a public road, within a town or village. The street is a jovial place where people do not cry, but mingle indiscriminately with others for business, social matters and politics. It would appear therefore that a street exists and has functions.

The functions of the street to provide access to property can be extended to include access to property by pedestrians and the employees of public utility services. Without the street, access to private or public property would be impossible. The access routes provided by the street also enable service providers to deliver the following to houses or other structures:

(a) Water
(b) Electricity
(c) Telephone
(d) Gas
(e) Sewerage
(f) Police surveillance
(g) Fire protection and circulation of fire-fighting equipment
(h) Storm drainage
(i) Mail drop and pickup services
(j) Socializing space (parks in central town)
(k) Meeting place (as an extension of the house)
(l) Business space (for stores, open markets, telephone booths, kiosks)
(m) Ambulances for emergency pickups
(n) Advertising space.

One of the functions of the street is to *provide light and air* to the buildings. Since we humans depend on air for our existence, the street plays the important role of bringing light and air to the buildings constructed on it. This is of particular importance in very populated urban areas like Nairobi, Abuja, Lagos, Douala and Yaoundé where high rise buildings line the streets. The light provides ventilation to the buildings, while the air brings life to both plants and human beings within the city (see PICs 6, 19).

Another function of the street is to control and channel the traffic desires of competing users: vehicles, pedestrians and motorcycles. It is difficult to imagine what would happen if vehicles, motor cycles and pedestrians were allowed to operate on a large unmarked space within an urban or city area. The uncomfortable result would be chaos and total confusion. Because of its demarcation, the city street system therefore allows the motor vehicle and motorcycles a one-dimensional degree of freedom in space (on a lane on a street).

It should be noted that the function of a street may change with time, depending on street uses as well as urban planning and zoning. However, its physical aspects may remain for centuries; examples are found in Rome, Moscow, Paris and London.

2.2 Classification of city streets

Once the functions of each street have been established, it must also be determined how the existing conditions cater for established functions, the level of development of the area, the topography, location of the urban area, and the characteristics of the users. When the street system is divided into four, five, or more discrete classifications, then each class must have specified standards and, by implication, serve and satisfy particular functions. The combination of functions to be performed by each class of streets is unique.

Chapter 2

Photo 1. Yaoundé street invaded by hawkers.

Photo 2. A market on the street.

Photo 3. Dual carriageway at Nlongkak, Yaoundé, improves traffic flow.

Photo 4. The street as marketplace.

The City Street in Camaroon

Photo 5. Lack of education and adequate training of drivers is causing traffic jams even on streets with adequate capacity

Photo 6. Street function – allow light and air into buildings, Yaoundé

Photo 7. Traffic congestion in Melen – Yaoundé, right lane going up to EMIA

Photo 8. U-Turn creates traffic disorder

It is recommended that the degree to which an existing street system satisfies the established classification should be reviewed from time to time. In this way, any conflict between users and classification can be seen at each review and hopefully corrected. It may be worthwhile for some of the money being invested in street modifications to be channelled into an in-depth study of the classification of existing streets and the functions they perform, before attempting to investigate the existing streets for meaningful short- and long-term modification or redesign. Additional research is essential to determine which of the functions currently being performed by existing streets are compatible with other urban activities.

A realistic organization of available urban space requires answers to pertinent questions. How compatible is the function of a street as an open market with vehicle traffic circulation? How and where should advertising billboards be placed? Are there rules governing the size and duration of billboards on the city street? These and other questions may find answers in the demand for space on the street.

Depending on the size of the urban area, streets can be classified into four systems: (i) the urban principal arterial system; (ii) the collector street system; (iii) the local street system; and (iv) transit routes. Each is described below with local examples.

The urban environment has a system of streets comprising the *urban principal arterial system*; it also includes some highways which can be identified as unusually significant to the area in which it lies in terms of the nature and composition of the traffic it serves.

The importance of a street is derived from the service provided to traffic passing through the area. The street that originates in the major roundabout on the old Douala Road leading to Nkolbison and that passes through Mvog-Betsi Market and continues to the Handicapped Centre, Etoug-Ebé Junction, TKC Junction, Mendong, and out of the Yaoundé city limits, constitutes an urban principal arterial system. The stretch from the Central Post Office Roundabout to Nlongkak Roundabout that continues to Bastos-Nlongkak, Etoudi, and Emana, is another arterial system.

These streets carry the major portion of traffic entering and leaving the urban area as well as the majority of vehicles desiring to bypass the city centre. In addition, significant intra-area travel, such as between central business districts and outlying residential areas, is also observed.

In practice, a properly designed and well-developed arterial system should help to define residential neighbourhoods, industrial zones and commercial areas. They should also minimize conflict between educational, health and other public service providers (schools, government ministries, banks).

A *collector street system* should include all distributor and collector streets serving traffic between major arterials and local streets and those streets used mainly for through traffic movements within a local area and for access to abutting property. From the Wouri Bridge through Bonabéri in Douala to Bekoko Junction is a good example of a major arterial and collector system.

The *local city street system* includes streets which are primarily used for direct access to residential, commercial, industrial and other abutting property. In Cameroon, however, there is no defining boundary between residential, commercial and industrial areas because the tendency is to mix residential and commercial activities (this is due to a lack of zoning laws).

The efficiency of any street system depends on the balance between the various types of street provided for within the urban setting – arterial, collector, and local streets. Growing traffic congestion within urban centres has made it very difficult for purely transit traffic to operate. Travelling from Edéa to the outskirts of Douala, a distance of about 62 km, may take between 40 and 45 minutes. Transiting Douala may take as long as two hours, depending on the time of travel. Transit routes should therefore be laid to minimize traffic congestion and provide maximum service to the community.

The difficulties we face today in the provision and management of infrastructure (water, electricity, roads, housing, organization of our towns, ports, airports, dams, telecommunication, information technology) exist because of a missing link. That link is discussed in the next chapter.

CHAPTER 3
The engineer and society

The engineer may be poorly paid, sometimes disregarded and ignored by society as has been the case here, but an engineer provides invaluable services to the community, including planning, design, research, transportation, electrification, maintenance, construction, and traffic management services. The engineer represents the biblical stone that was rejected but became the cornerstone. We need to understand the roles of engineers in order to appreciate them. However, before we do so, it may be necessary to find out what contributions the engineer can make towards the functional organization of the city state. Some engineers have the eyes of eagles and the innovative minds of bees. This is a tribute to those engineers who have changed the well-being of humanity through their work.

In 1868 John Seymour Lucas, during the conference of engineers at Menai Strait, Wales, prepared an outline for the occupation. According to Lucas, an engineer is a professional in the applied sciences. Engineering competencies include mathematics, scientific knowledge, and management skills. Fields of employment for engineers include research and development, industry and business, science, architecture, and project management.

An engineer is a professional practitioner of engineering concerned with applying scientific knowledge, mathematics and ingenuity to develop solutions to technical and practical problems. Engineers design materials, structures, machines and systems while considering the limitations imposed by practicality, safety and cost.[3,4] Further, the word "engineer" is derived from the Latin root ingenium, meaning "cleverness".[5]

Engineers, then, are grounded in applied sciences, and their work in research and development is distinct from the basic research focus of scientists. The work of engineers forms the link between scientific discoveries and the applications that meet the needs of society.[6]

---

[3] Bureau of Labor Statistics, U.S. Department of Labor (2006) engineers.
[4] National Society of Professional Engineers (2006).
[5] *Concise Oxford English Dictionary*, 1995.
[6] Occupational outlook Handbook, 2006 – 07 Edition.

3.1 Roles and expertise

Engineers develop new technological solutions. During the engineering design process, the responsibilities of the engineer may include defining problems, conducting and narrowing research, analyzing criteria, finding and analyzing solutions, and making decisions. Much of an engineer's time is spent on researching, locating, applying, and transferring information.[7]

In continental Europe, Latin America, Turkey and elsewhere, the title is limited by law to people with an engineering degree, and the use of the title by others is illegal. In the United Kingdom and the United States, the title is limited to people who have an engineering degree and have passed a professional qualification examination. In Egypt, the educational system makes engineering the second most respected profession in the country after medicine. Engineering colleges in Egyptian universities require extremely high marks on the General Certificate of Secondary Education. In Cameroon, the practice of the profession of civil engineering is regulated by Law No. 2000/09 of 13 July 2000.

Sometimes the public tends to confuse engineers with other professionals; this is unfortunate. Engineers bring their scientific know-how to assist society in solving problems. For example, now that the world is in danger from climate change, we have engineers watching the weather; climate change has become a hot topic for environmental engineers. They are working on alternative sources of energy, renewable energy using natural resources to generate wind, solar, hydroelectric, and geothermal energy.

It is essential to emphasize that creativity is inherent in engineering. Creativity is not just about being artistic; it is about having the technology and the creative mind to know how to apply technology to design so that the end product is aesthetically pleasing and user-friendly. Solving problems is the best part of engineering practice. The Soviet engineers who conceived, designed and built the first manned sputnik and later the first space station were visionaries; they applied creativity to resolve the practical problems of space research. Their initiative opened the way for America to land a man on the moon in 1969.

Practical problem-solving is what drives engineers forward. It does not matter what the problem is; an engineer will find a solution, whether it is in surveying, designing, estimating quantities, or constructing a road. For example, the author's engineering team provided a solution to the challenge facing land transportation to replace the use of ferries to move goods and people from Ekondo-Titi to Mundemba in Ndian Division.

The road transformed people's lives in the area by providing land access where previously there was only access by sea. The journey took a grueling 12 hours through the creeks, with occasional accidents and loss of lives and property. On completion of the road, the journey time between these localities was reduced to two hours. The solution to the problem was made possible by the ingenuity of Cameroonian engineers (Tamajong Ndumu, Omar B Sendze, Jerome Obi Eta, Michael Ajua Alemanji, and Michael Takoh).

---

[7] A. Eide, R. Jenison, L. Mashaw, L. Northup. *Engineering: Fundamentals and Problem-solving*. New York: Mc Graw Hill, 2002.

Science is knowledge based on observed facts (experimentation), and tested truths arranged in an orderly system that can be validated and communicated to other people. Engineering is the creative application of scientific principles used to plan, direct, build, guide, manage, or work on systems to maintain and improve our daily lives. But why have engineers been relegated to the back benches in our communities? One of the main reasons is that public memory is short. For example, when the Mungo Bridge collapsed, it was to the engineers that society turned for a solution. It was they who provided the temporary crossing and later the permanent bridge, but once the bridge was put in place, people forgot who had executed the job.

Perhaps of greater significance in the decreasing public awareness of the profession is the fact that average people do not have any personal dealings with engineers, even though they benefit from their work every day by the use of highways, electricity, water, streets, hospital buildings, telephones, television, court houses, ministerial buildings, markets. By contrast, people visit a doctor at least once a year for health matters, the accountant for taxes, the teacher, and occasionally even a lawyer.

An example of the side-lining of Cameroonian engineers occurred in 1997. On 14 September 1997, the breaking news from Cameroon Radio and Television (CRTV) was the collapse of a six-storey building in the Dakar District of Yaoundé, near Nsam Junction (about 150 meters from the SCDP petroleum depot towards Ahala Village on the Yaoundé-Douala Highway) (see PIC 35).

The Cameroon Society of Engineers quickly designated a team of three specialists (a structural engineer, a materials expert, and a geotechnical expert). The team carried out a comprehensive technical inquiry and published a scientific report which discounted an explosion or any foundation failure.

The causes of the collapse were judiciously traced to the poor design of the structure and the total absence of any supervision of the construction by qualified engineers. It was unearthed that extensive deterioration had occurred leading to carbonation and corrosion of the reinforcement as a result of the unfinished nature of the structure at collapse (over the years).

Specifically, it was determined that low concrete densities resulted from various causes: poor aggregate, high water ratio, segregation of aggregate during concreting. Lime leaching from exposure to the elements contributed significantly to system failure. Construction had been on-and-off for 18 years without adequate protection of the infrastructure from the forces of nature (cf. *A Technical Inquiry Commissioned and Financed by the Cameroon Society of Engineers April 1998* by Shutang Mungwa, MCSE; U Chinje Melo, MCSE; and Paul Tamajong, FCSE).

This report was duly forwarded to the competent authorities with detailed recommendations. No reply or feedback was ever returned to the Society. New cases of buildings collapsing in our cities (cf. the six-storey building in Mermoz Street, Akwa-Douala, the one-storey building in Afrique du Sud, Bonamoussadi District in Douala, and a building at Mobil-Guinness, Ndokoti-Douala) point to the non-adoption of the recommendations and a tacit confirmation that the report was never read.

This is a typical case of a prophet not accepted in his own country. In other countries, the proposals would have led to the establishment of local laws to deal with public safety with regard to the construction and maintenance of high-rise structures. Such laws enhance the value of construction and improve city landscapes. These incidents demonstrate a complete misunderstanding of, and basic ignorance about, the role of the engineer in society.

3.2 Facing up to the dilemma

The mission of engineers is to transform the face of the earth, conceive, design, build, maintain and control the putting in place of infrastructure of all kinds for the advancement of mankind. They apply creativity to resolve the practical problems of societies.

Although engineers may be poorly paid, very often ignored, and sometimes even abused, the truth is that without them, society could not function. Engineers use their ingenuity to transform ideas into projects which result in infrastructure, water and sanitation systems, information technology, and countless other advances.

Without the input of engineers, life in society would be unbearable. Imagine the following brief scenario. An electricity supply failure leads to the shutdown of pumps at the water supply station, leaving towns without this vital utility. In hospitals, the X-ray machines cannot function, no laboratory testing can be done, and doctors cannot fully treat patients. Then the engineer appears and uses expertise to resolve the problem. However, there will be no TV or radio announcement about the fact that engineers solved the problem.

This persistent neglect of the role of the engineer has led to improvization by those who think they can act as engineers without having the required rigorous education and professional training. The results stare us in the face daily: kerb parking in front of traffic lights (marked out by workers from the city government whose only qualifications are a brush and paint) and the absence of directional sign postings in cities. These are indicators that engineers have not been involved in the operations.

The complete absence of traffic engineers within the staff of city governments is a monumental hindrance to urban development in Cameroon. The sections that follow detail the importance and significance of traffic engineering in urban and regional planning.

CHAPTER 4
Traffic engineering

Fortunately, science is universal, unlike politics with its subjective thinking. Some nations have claimed the invention of their own model of democracy different from the Greek model established centuries ago as a government of the people, for the people, and by the people.

The American practice of traffic engineering is worthy of emulation for various reasons. Mass production of the automobile began there in the early 1900s; by the 1930s the country was the largest automobile manufacturer in the world. It is therefore safe to assume that cars, trucks and other vehicles have been on American highways and streets for over 100 years, an invaluable experience has been gained. This author's personal travel experiences around the world (Moscow, St Petersburg, Kiev, Rome, Milan, Paris, London, Berlin, Frankfurt, Hamburg, München, Brussels, Amsterdam, Tashkent, Cairo, Riyadh, Jeddah, Lagos, Accra, Freetown, Kampala, Nairobi, Addis Ababa, Mombasa, Durban, Johannesburg, Charlotte (NC), Los Angeles, San Diego, New York, Washington D.C., Virginia, Philadelphia, Boston, Delaware, Toronto, Ottawa, Montreal, Quebec) have led him to the conclusion that America has the most organized streets.

It is therefore reasonable to look in some detail at the practice of traffic engineering in America. According to the Institute of Traffic Engineers, traffic engineering is "that phase of engineering which deals with the planning and geometric design of streets, highways, and abutting lands, and with traffic operations thereon, as their use is related to the safe, convenient and economic transportation of persons and goods" (cf. *Highway Engineering Handbook*, p. 3).

4.1 The traffic problem

Traffic problems arise from the inability of society to match demand for street and highway facilities with growth in traffic volume. Congestion and delays are the direct outcome of the disequilibrium between management of the available space and the ever-growing traffic volumes registered on our streets. As a result, more and more accidents are being reported on streets and highways.

Traffic engineering attacks the problem of traffic accidents and congestion from two approaches: (a) the constructive approach and (b) the restrictive approach. The constructive approach includes the planning and geometric design of new streets, highways, transit systems, and parking facilities to meet estimated future needs for transportation and termination.

The restrictive approach implies the maximization of the efficiency of existing streets and highways through the application of traffic regulations and traffic control devices. Controls place restrictions on the driver's freedom. Traffic studies and analyses provide the basis for both the constructive and restrictive approaches.

4.2 Traffic regulation

Traffic regulations provide the controls which are required for efficient use of streets and highways. Some controls have area-wide applications (such as driving on the right), whereas others are applied at specific locations (such as one-way control or parking prohibitions). Practically all regulations are based on state laws and ordinances.

The state has the main responsibility for providing the legal basis for traffic control through the state motor vehicle code. This code generally provides the following:

(a) General rules on driving and walking applicable to all public highways (speed limits, right-of-way, U- turns, meaning of signals);

(b) Rules on accident reporting, vehicle equipment, licensing of drivers, financial responsibility, vehicle registration;

(c) Enabling legislation to permit local authorities to adopt regulations applicable to specific locations (speed zones, one-way streets, parking control, turn prohibitions).

The Uniform Vehicle Code (29) is a guide for states to use in adopting uniform traffic laws. The Model Traffic Ordinance (30) guides cities in preparing traffic ordinances. Universal adoption by states of uniform general rules for driving and walking reduces the chances of misinterpretation of the law and the resulting confusion, delays, hazards, and possible accidents.

The Model Traffic Ordinance supplements the state-law provisions by providing the following:

(a) Regulations applying throughout the city, such as pedestrian control, bicycle control, parking control, and emergency vehicle control;

(b) Schedules of special regulations applied to specific locations (through- streets, one-way streets, parking prohibitions and time limits, speed zones, stop signs, yield control, and traffic signals);

(c) Administrative provisions establishing the position of traffic engineer, the police traffic division, the traffic-violations bureau, and delegation of authority to the traffic engineer (cf. *Highway Engineering Handbook*, Section 7, Donald S. Berry).

Traffic engineers are the nerve centre of the technical expertise in town and city councils or governments. Their absence in Cameroon is felt everywhere. For example, there are no defined traffic zones, no speed limits on streets, very few traffic lights at junctions, no planning of towns and cities, and no street maps. This is an unfortunate vacuum that must be filled urgently.

Furthermore, our current street system was designed without any provision for kerbside parking, bus stops and, in some places, sidewalks for pedestrians. The vehicle boom has reduced the capacity of the streets, forced away pedestrians who walked on the same level as the vehicles, and brought about the need for kerbside parking, and adequate sidewalks.

Redressing the above disorder would require the expertise and input of traffic engineers. Adequate exploitation of our city streets requires upgrading and the effective use of traffic engineering coupled with effective policing of traffic regulations in force.

Some of the current congestion can be mitigated by the simple use of "NO LEFT TURN" signs placed at appropriate locations at road junctions, creating more reasonable parking restrictions, additional traffic lights and improved lane markings, etc. For example, the small roundabout in front of Pharmacie du Soleil should have two "NO LEFT TURN" signals: one for the traffic ascending the hill from the Hilton Roundabout to prevent traffic drifting to Carrefour Wada, and the other for traffic flowing downwards from the junction of Direction des Impots, to stop drivers driving to Marche Central and blocking the smooth flow of traffic because they should have done so using the street that passes by Bricolux. In almost all our major cities in the country, the quality of traffic engineering services is poor, or simply not available. The establishment of one-way streets where there are parallel streets would ease congestion.

This author is convinced that the allocation of modest expenditure to traffic engineering, together with the widening of certain intersections, could reap proportionally larger benefits without greatly infringing upon the liberty of individual road users.

The enforcement of the existing parking regulations in Yaoundé is vigorous, but "no-parking" zones must be clearly indicated to avoid the embarrassment faced by road users whose vehicles are immobilized when they park on non-identified spots only to be told verbally that it is a no-parking area. (Such practice is illegal and must be stopped.) The regulation of parking and other traffic activities must become the subject of an in-depth study leading to the enactment of laws governing the use of city streets and other public places. Improvement of traffic flow is a business opportunity that can generate enormous revenue for the city council or government.

The most effective way of enforcement is by upgrading the quality and number of personnel involved. This would certainly require more resources, but part of the congestion levies should be directed towards this end. Higher salaries for enforcement officers would help prevent the once common use of "unofficial transfers" as a means of avoiding legal penalties.

Improvement in personnel and modification of administrative procedures will not solve all of the problems of enforcement. However, steps must be taken to overhaul the rather archaic and sometimes non-existent legal system. The present legal structure was not designed to accommodate a modern system of traffic enforcement.

The levying of fines for traffic infractions should be an almost automatic procedure with a specific penalty associated with a given offence. Traffic violations should be charged to court, where the penalties are confirmed and the associated fines paid. This is in contrast to the present spasmodic system where the penalty is debatable and varies depending on who you are and what language you speak. Special courts should be set up to handle traffic cases so as to discourage the excessive use of an already overburdened judicial system.

The duration of kerbside parking allowed legally must be clearly marked and appropriate signs posted for all potential users to see. Parking information and regulations should be adequately published and disseminated in all of the media. Current horizontal markings must be completely reviewed because most of them are inappropriate.

When traffic regulations have been widely circulated and the public educated, the next phase is enforcement. This requires the application of defined penalties. Those who thereafter refuse to comply can and should be punished for willingly violating the law.

CHAPTER 5
Why streets should be planned

Some experts might speculate about how the street, as an innovative research subject, should be approached scientifically. Planners can use new technological tools of analysis based on accumulated data and other research findings to indicate the effect street development will have on economic growth and future traffic needs.

Streets have functions and classification and thus require advanced planning. In the past, the city street was disregarded, but the complexity and nature of activities witnessed on our streets invite us to a more rigorous and disciplined thought process. Consequently, it is imperative that future comprehensive transport planning policies for our towns should be oriented towards ensuring equal opportunities for motorists, pedestrians and those who have no desire to become completely dependent on cars. This would allow and encourage people to settle in places that are accessible to normal amenities and still benefit from reliable public transport like buses.

Visible changes are currently taking place in our towns as increasing affluence and better transport facilities allow people to satisfy their demand for more space as well as more pleasant and friendly surrounds on the fringes of urban centres. These locational changes occur constantly as towns evolve. Spontaneous growth can be observed in new population centres like Odza, Nkomo, Awae in the Yaoundé neighbourhood; Nylon Village and parts of Bonabéri in Douala; Foncha Street-Nkwen in Bamenda; Mile 4 in Victoria; and the surrounding areas of Alaska Street in Buea Road, Kumba. These should not be allowed to evolve unplanned. These changes should be reflected in well- coordinated planning regulations, the supply price of land, and in the provision of appropriate and flexible transport facilities as well as water, electricity and communications. This would avoid repeating the errors of the past which have led to the present chaos. An efficiently planned and managed public transport system should control the development or expansion of urban settlements, thus promoting greater use of existing infrastructure and allowing more time for necessary adaptation.

Human activity (that is, where people live, work, shop, go to school, and socialize) is an important factor affecting the provision of transportation. The street seems to have become the magnet for these activities and therefore should become an integral part of town planning. Such planning includes gathering, analyzing and recording pertinent data. The following data should be collected: an inventory of the historical records on existing streets; the functions and classifications of existing streets; the physical state of streets with regard to traffic circulation and maintenance needs; types of traffic on streets; accidents; existing legislative records; and environmental regulations.

The specific objectives and goals of street planning are to:

(a) Provide the city and town councils and governments with robust recommendations for short- and medium-term plans for integrated development that incorporates the city street as the central element of travel;

(b) Provide a basis for educating the population on the proper use of streets;

(c) Assign proper names to streets, junctions, roundabouts and monuments as well as identifiable numbers to houses and other buildings;

(d) Facilitate the creation of viable street maps of towns and include these in the Global Positioning System;

(e) Provide a scientific basis for the prioritization of project activities, their budgeting, and subsequent realization;

(f) Ensure the complete coverage and adequate distribution of activities, including well-defined job descriptions (for example, responsibilities for preventative street maintenance, control of no-parking zones or restitution of traffic signs);

(g) Ensure the permanent functioning of all drainage structures and systems through regular inspection and timely technical interventions.

5.1 The consequences of failure to plan

Without coordinated plans of activities to guide them, council management may become complacent and ineffective in the execution of their duties, and wasteful in the use of resources. The problems of ignorance about what to do and when to do it are legend, and we cite only a few cases to illustrate the magnitude of the planning problem.

The use of ditches purely as drainage structures has been widely abused in our cities and towns. Ditches have been transformed into refuse dumps; they quickly fill up with debris and stagnant surface water. The transition to permanently covered ditches has also not worked well because the cover slabs are poorly designed and are frequently broken by heavy trucks. Broken slabs then fall into the ditches, are abandoned, and obstruct the normal flow of water.

Traffic signs are knocked down by reckless motorists but are not quickly repaired or replaced, leaving a potentially dangerous information vacuum. Interdiction signs are not always respected. Zebra crossings are marked horizontally on the street surface, but there are no vertical signs to signal their presence to motorists.

It is most unfortunate that the absence of planning has led to the misplacement of priorities as witnessed in our cities and towns. Instead of providing for the installation of traffic lights at congested junctions to improve traffic flow, council authorities are busy building market shops in excess of demand (as in Yaoundé for example) and completing other misdirected projects. In Bamenda, parking meters were installed despite the sporadic nature of the electricity supply. Most of the meters have been removed or stolen; many are simply standing idle. Manual collection of parking fees would have provided employment and perhaps been more effective, thus guaranteeing a return on the investment in meters.

Some progress has been made in the attempt to improve traffic circulation. At locations where interdiction signs have been posted, the fines for non-compliance have also been indicated. For example, at the roundabout where Commercial Avenue begins, there is a stop sign which also carries a fine of 25,000 CFA francs for non-compliance. Because there are no enforcement measures put in place to guarantee compliance, taxis stop there to pick up and drop off passengers. During peak hours, this causes severe traffic congestion at the roundabout.

It has already been emphasized that actions taken to improve traffic circulation should be backed up by education of motorists and the general public. Enforcement should then be put in place to ensure compliance. In Kumba, there was an advanced procedure of street naming about five years ago. Unfortunately, these street sign postings are disappearing, and no one seems to be interested in re-establishing them or continuing the street naming exercise. The town of Mamfe has some of the best paved streets, but traffic circulation is hindered by the irrational placing of speed brakes on the street. In certain places, there are as many as three speed brakes within a short distance of less than 200 metres. Speed brakes are meant to slow down vehicle movements, but it is impossible to gain speed within such a short distance. This is a clear example of lack of planning and technical expertise.

5.2 *The impact of lack of planning*

Adequately planned and efficiently executed projects usually guarantee returns on the investments made; in the worst cases, the projects break even. Conversely, the absence of preliminary planning may derail the achievement of the goals for which the projects were conceived, designed and implemented (to render a housing service or medical care in these cases). Because urban transportation is dynamic, any project that has a transport component must also have a plan that monitors traffic volume as an indicator for updating infrastructure planning.

On 12 December 1985, during the examination of bid offers by the National Tenders Commission, it was suggested that instead of just constructing streets within the vicinity of housing complexes (thousands of housing units were under construction), the programme should be extended to cover the provision of a dual carriageway. This street would link the houses being built in Akwa-North in Douala with the Bonabéri Roundabout in the City centre. The proposal was rejected as too sophisticated; because the members of the commission were all administrators, they failed to grasp the link between providing adequate transport infrastructure to a housing estate and its future profitability.

Providing proper transport access to the housing units would attract customers to hire or purchase these houses. A good access road should therefore be part of the project costing or budget; without an access road, no customers would opt to hire the units. The project was completed and commissioned nearly two decades ago, but the occupancy rate of this vast project that swallowed billions of francs of taxpayer money has never reached 70%. The reason for this is the perennial traffic congestion on the street(s) leading up to the housing complex (which has since been surrounded by other real estate developments and social infrastructure). It is troubling that no one has tried to resolve this transportation problem for the people who are trapped in these houses without adequate access to the city.

Two other projects deserve to be mentioned as victims of the failure to plan. The access roads to the Douala Referral Hospital and the General Hospital in Yaoundé were constructed after 1986. Unfortunately, they were not based on adequate origin and destination traffic studies within the neighbourhoods of these hospitals. Also, there were no projections made for new settlements or future population growth. These omissions were monumental mistakes of project planning that have come back to haunt the owners financially. Unfortunately, the construction of these access roads opened up the land, drawing into the area thousands of people that were looking for cheap land on which to build their houses.

These streets that were once effective ways of reaching the hospitals have become very congested. Because rapid and effective transport to the hospital can make the difference between life and death, many people now fear taking patients there, especially those who need emergency treatment. Unless these access roads can be upgraded to handle current and future traffic projections, it will be difficult for these hospitals to ever break even financially. Although they are currently sustained by subventions from the state because they are health institutions, their economic states are unstable.

The most devastating negative impact of lack of planning is the increasing number of abandoned buildings littering the landscape of our cities. These and other ad hoc cases provide mounting evidence illustrating the degree of misplaced priorities and waste of resources encountered by our city and town councils. There is therefore a compelling case in favour of sustained urban street planning. Further research on the merits of street planning would be beneficial. And to provide a wider perspective, it may not be superfluous to collect data on the national vehicle fleet.

CHAPTER 6
Evolution of the Cameroon motor vehicle fleet, 1980 to 2009

Any research work on the street which fails to take cognizance of the national vehicle fleet ought to be discarded as inappropriate. The street represents the most circulated infrastructure in most countries of the world (except of course in those countries with underground transport systems). In economic terms, the street can be said to represent the critical path of economic development. The street is the principal home of vehicles, either during the day or at night when their owners are at rest. Vehicles that have gone out of service or are broken down are often abandoned on the street to become eye sores. The size of the vehicle fleet affects pollution, degree of traffic congestion, space occupation and environmental degradation.

6.1 Growth in vehicle numbers

Our research findings cover the period between 1980 and 2009 and correspond with the era of recorded statistics. The data show that the highest growth rate of 24.3% occurred between 1986 and 1987 (Table 6). However, if we exclude the intermittent negative growths registered for six years, the average growth rate recorded would be 7%.

Table 4: Number of vehicles in Cameroon, 1980 to 1986

Year	Cars	Trucks, pickups	Buses	Trailers, semi-trailers	Road tractors	Motorcycles	Total
1980	45,000	23,400	3,700	1,600	1,600	30,200	105,500
1981	47,800	23,400	3,800	1,600	1,600	36,400	114,600
1982	53,300	25,600	3,900	1,600	1,600	43,500	129,500
1983	57,600	25,300	4,400	1,600	2,200	51,700	142,800
1984	64,200	24,800	3,800	1,700	2,300	55,900	152,700
1985	75,000	25,100	4,100	2,000	2,300	58,200	166,700
1986	88,800	25,900	5,000	2,300	2,700	58,700	183,400

Source: Ministry of Public Works and Transport, Statistics Office.

Table 2: Number of vehicles in Cameroon, 1987 to 1991

Year	Cars, trucks, pickups		Buses	Trailers, semi-trailers	Road tractors	Motorcycles	Total
1987*	188		220	1,087	1,356	37,300	227,963
1988*	186		246	1,193	1,342	26,335	215,116
1989*	173		245	1,194	1,284	20,758	196,481
1990	105,900	35,400	6,800	3,000	3,300	44,500	198,900
1991	104,600	32,500	6,000	3,100	3,400	43,800	193,400

Source: Ministry of Transport/Department of Land Transport.
*The recording system was modified between 1987 to 1989 and vehicles were registered not by type but by mark, except for trailers, road tractors and motorcycles.

It was also observed that the vehicle fleet peaked in 1987 with a registered number of vehicles reaching 227,963 (Table 2). It is very likely that this may correspond to the change in policy from importing only new vehicles, to bringing in anything that moves (Tables 3, 4 and 5).

The histogram in Table 7 depicts a declining fleet from 1988 to 1994. Negative growth was even registered (Table 6). This trend may in turn reflect the impact of the economic crisis witnessed in Cameroon during the late 1980s. Growth returned timidly in 1995. It picked up in 2000 and has grown steeply ever since, with attendant consequences on street infrastructure.

The age factor was captured in the 2000 to 2009 statistics where the vehicles were registered by defined age brackets. The age of vehicles has a three-fold impact on the street. Because old vehicles are slow-moving within the traffic stream, they create congestion on the street. Congestion leads to pollution and environmental degradation. Those vehicles no longer in service occupy needed space when abandoned on the street.

Table 3: Number of vehicles in Cameroon, 1992 to 1999

Years	Cars	Lorries, vans	Trucks	Buses	Trailers	Tractors, machines	Cycles	Total	Growth ratio %
1992	98,100	19,000	11,800	5,500	3,100	3,500	42,800	183,000	
1993	92,800	17,200	11,800	5,000	3,000	2,000	40,000	171,800	-6.52
1994	88,300	16,000	10,500	4,600	3,000	3,000	39,500	164,900	-4.02
1995	94,757	16,239	10,629	7,158	2,554	3,048	40,003	174,388	5.75
1996	100,939	16,566	10,814	7,366	2,723	3,175	42,838	184,421	5.75
1997	102,248	16,756	11,002	7,586	2,895	3,385	44,492	188,365	2.13
1998	105,865	17,446	11,118	8,088	3,070	3,538	44,801	193,926	2.87
1999	110,777	17,827	11,278	9,313	3,087	3,718	45,948	201,948	4.13

Source: Ministry of Transport, Department of Land Transport.

Table 4: Number of vehicles in Cameroon, 2000 to 2009

Years	Cars	Trucks, pickups	Buses	Trailers, semi-trailers	Tractors	Motorcycles	Total fleet	% of fleet Motorcycles	Vehicles
2000	115,917	29,631	10,725	3,106	3,909	46,987	210,275	22.35	77.65
2001	134,507	32,075	11,734	3,191	4,102	47,419	233,028	20.35	79.65
2002	151,853	35,255	13,758	3,220	4,680	52,912	261,678	20.22	79.78
2003	173,137	35,490	13,908	3,250	4,723	53,319	283,827	18.79	81.21
2004	164,429	40,577	14,937	4,133	5,201	62,517	291,794	21.43	78.57
2005	175,981	43,417	15,982	4,422	5,564	66,893	312,259	21.42	78.58
2006	183,020	44,064	16,622	4,598	5,784	69,569	323,657	21.50	78.50
2007	188,985	46,134	17,164	4,692	5,902	71,863	334,713	21.46	78.54
2008	195,077	47,132	17,659	4,763	5,991	73,908	344,531	21.45	78.55
2009	200,753	47,770	18,030	4,797	6,034	75,460	352,844	21.39	78.61

Source: Ministry of Transport, Cameroon. Transport Statistics Bulletin, 2010 edition, p. 28.

Table 5: Cameroon vehicle fleet, 1980 to 1986 and 2000 to 2009

Vehicle type	1980 to 1986	Recorded increase (%)	2000 to 2009	Recorded increase (%)
Cars	45,000 to 88,8000	43,800 (97)	115,917 to 200,753	84,835 (73)
Trucks, pickups	23,400 to 25,900	2,500 (10.7)	2,9631 to 4,7770	18,138 (61)
Motorcycles	30,200 to 58,700	28,500 (94)	46,987 to 75,460	28,473 (60.6)

The upward trend in the number of cars and motorcycles is alarming. Within a short period of 20 years (1986 to 2006), private cars more than doubled, climbing from 88,800 to 183,020 (94,220) units, an increase of 106%. Trucks and pickups grew from 25,900 to 44,064, adding 18,164 units or 70%; motorcycles 58,700 to 69,569, adding 10,869 units or 18.5%.

Table 6: Evolution of Cameroon vehicle fleet showing growth rates (%)

No.	Year	Total vehicles	Growth rate %*
1	1980	105,500	
2	1981	114,600	8.6
3	1982	129,500	13
4	1983	142,800	10.3
5	1984	152,700	6.9
6	1985	166,700	9.2
7	1986	183,400	10.03
8	1987	227,963	24.3
9	1988	215,116	-5.62
10	1989	196,481	-8.71
11	1990	198,900	1.2
12	1991	193,400	-2.76
13	1992	183,800	-0.05
14	1993	171,800	-6.52
15	1994	164,900	-4.02
16	1995	174,988	6.11
17	1996	184,421	5.39
18	1997	188,365	2.13
19	1998	193,926	2.92
20	1999	201,948	4.14
21	2000	210,275	4.12
22	2001	233,028	10.82
23	2002	261,678	12.29
24	2003	283,827	8.46
25	2004	291,794	2.81
26	2005	312,259	7.01
27	2006	323,657	9.75
28	2007	334,713	3.42
29	2008	344,531	0.24
30	2009	352,844	2.41

*Created from Tables 1 to 4 of the statistics. *Some growth rates were negative. The average growth rate of 7% excludes the six years of negative statistics recorded.*

Table 7: Total vehicles registered by year

No.	Year	Total vehicles
1	1980	105,500
2	1981	114,600
3	1982	129,500
4	1983	142,800
5	1984	152,700
6	1985	166,700
7	1986	183,400
8	1987	227,963
9	1988	215,116
10	1989	196,481
11	1990	198,900
12	1991	193,400
13	1992	183,800
14	1993	171,800
15	1994	164,900
16	1995	174,988
17	1996	184,421
18	1997	188 365
19	1998	193,926
20	1999	201,948
21	2000	210,275
22	2001	233,028
23	2002	261,678
24	2003	283,827
25	2004	291,794
26	2005	312,59
27	2006	323,657
28	2007	334,713
29	2008	344,531

Source: Crafted from recorded statistics of the Ministry of Transport and Public Works.

Diagram 2: Histogram of registered vehicles in Camaroon*

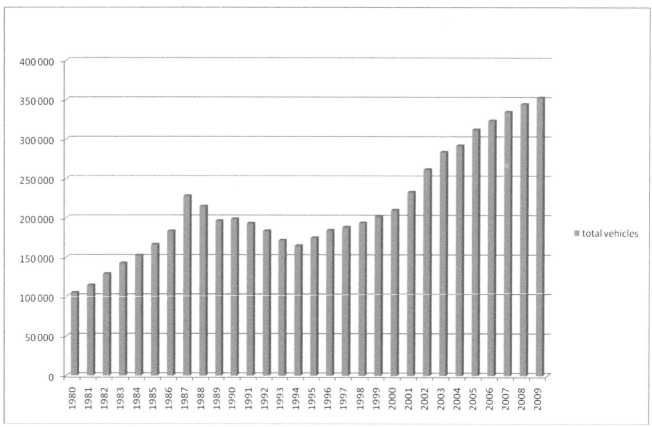

Designed from statistics in Table 6.

There are those who argue in favour of the importation of vehicles regardless of age; they cite the increased government revenue from custom duties and user taxes of all kinds to defend this preference. However, they should be requested to update the community on the costs of pollution, environmental degradation, health problems as a result of air pollution, traffic congestion, and expanding street infrastructure to accommodate the perennial increases in the vehicle fleet. For instance, the number of cars registered in 1980 was 45,000, and the number of motorcycles was 30,200. In 2009, the number of cars escalated to 200,753, an increase of 155,753, or an incredible jump of 346%. This is unprecedented growth by any standard. Motorcycles increased to 75,460, a jump of 149.89%

    A meticulous review of the motorcycle sector has shown that their numbers almost tripled within the last 20 years. About fifty motorcyclists were interviewed for this book, and it was discovered that some of them register their motorcycles, but a majority do not. The same vehicle operators declined to comment on issues of training, licensing, insurance and security. Based on statistics for 2009, motorcycles represented 21.39% of the national vehicle fleet. Consequently, 75,460 registered operators had not been taught the Highway Code, thus setting them up for havoc in the flow of traffic.

We were also informed that, on average, each motorcyclist takes home 6000 CFA francs per day. Assuming that they charge 200 francs per drop, each motorcyclist transports 30 passengers per day, meaning a total of 2,263,800 passengers. Society thus entrusts millions of people into the hands of those who have not been schooled in the Highway Code, nor have they received any driver training. What does this unprecedented risk of death mean to society? Motorcyclists' propensity for recklessness is legend. A whole ward in Laquintinie Hospital in Douala has been named after them because it regularly receives victims of motorcycle accidents.

How can this sector be regulated, and how can the problem of too many vehicles crowding onto narrow highways and streets be resolved? Rational thinking people would immediately take restrictive actions to contain the perennial increase in the vehicle fleet. Restricting importation of vehicles to between the ages of 1 and 5 years would have an impact. Kerbside parking seems to compound the problem by narrowing the street and creating traffic congestion. A short-term measure might be to remove kerbside parking solutions adopted by councils. It might be more beneficial to widen the streets instead.

Table 8: Number of Vehicles in Camaroon by vehicle age and region, 2009

Age range (years)	Regions								
	Littoral	West	Far North	North	Adamawa	East	Centre	South-west	North-west
Less than 1	8,845	4,992	3,351	2,264	1,151	1,702	5,734	1,769	1,735
1 to 4	6,271	4,292	1,961	2,263	1,191	164	6,150	2,301	1,559
5 to 14	6,358	380	806	530	719	232	8,781	593	564
15 to 19	6,356	781	516	377	548	190	10,228	1,511	1,134
20 to 29	3,732	1,015	482	371	441	203	7,253	1,004	848
30 years+	144	51	49	24	50	4	214	36	32
TOTAL	31,706	11,436	7,163	5,829	4,100	2,494	38,360	7,348	5,872

Derived from Table 2.

Table 9: Vehicle fleet in Camaroon by vehicle age and region, 2009 (in %)

Age range (years)	Far North	North	Adamawa	East	Centre	South-west	Littoral	West	North-west	National average (%)
1 to 4	7.99	13.28	12.36	6.56	8.13	4.0	18.90	4.60	1.80	8.62
5 to 14	37.22	30.03	33.97	33.59	30.33	21.14	30.92	14.60	21.10	28.10
15 to 30+	54.79	56.69	53.67	59.85	61.4	74.86	50.18	80.80	77.10	63.28

Source: Derived from 2009 registration statistics.

6.2 Lessons learnt from vehicle fleet analysis

Because transportation plays an important role in the economic development of every nation, governments use a combination of policy measures, regulations and technical solutions to receive optimal benefits from the sector. Early research by consulting engineers offered numerous recommendations in their final report ("A Study of the Transportation Plan in Cameroon", February 1986). These are discussed below.

One recommendation was the wearing of safety belts by drivers and passengers, and the wearing of helmets by operators and passengers on motorcycles and motorized bicycles. In short, city authorities should regulate the motorcycle sector. Also, vehicles that have been involved in serious accidents and have been repaired should undergo compulsory inspection by qualified transportation personnel before they are allowed back into circulation. These recommendations have not been implemented.

Another recommendation was the withdrawal of driving licences and commercial licences in instances of verified responsibility for an accident causing death of a person independently of penal sanctions. In essence, the study called for the development of a basic code of regulations.

The analysis confirmed the mounting pressure being exerted on street infrastructure by the constantly increasing numbers of vehicles in the fleet. Therefore, the policy of open-ended importation of vehicles should be reviewed to apply age limit restrictions (1 to 5 years only). An in-depth examination should be carried out to find an adequate solution to the problem of traffic congestion (for example, street widening). It has become impossible to provide adequate space both on streets and on roads for the ever-growing number of vehicles registered every year. Table 9 illustrates that in 2009, only 8.62% of the fleet was between 1 and 4 years, and the majority (63.28%) of the vehicles were 15 to 30 years of age or older. This poses a problem of acute pollution, unavoidable congestion in certain streets and environmental degradation.

Development planners rely on robust statistical data for their plans. Unfortunately, this study of the national vehicle fleet from 1980 to 2009 revealed many loop-holes in the recording of statistical data. It was discovered that it was only in 2009 that vehicle registrations were done by category and vehicle age. Also, it was not possible to determine the cumulative national vehicle fleet at the end of each year. This was due to the fact that no records existed showing vehicles that had gone out of service for any reason: old age, unsustainable repairs, fatal accidents, or theft. This is very unfortunate because it renders the vehicle fleet statistics incomplete, and therefore, almost useless for planning purposes.

The constant growth of the national vehicle fleet imposes congestion and the urgent need to increase the size of some major streets (through the construction of additional lanes). Regular traffic counts must be undertaken to determine construction priorities.

CHAPTER 7
Demand for space in city streets

The street is the nerve centre of all activities in the city and therefore has become a pole of attraction. There is fierce competition among the rival forces vying for every square inch of street land. Because the land available is limited, the price must determine who will acquire land, especially in the central parts of the city where almost all available space has been occupied. This difficulty arises because no planning initiatives were put in place to guide the development of our cities and towns. They therefore developed on the principle of "spontaneous settlement" rather than spatial necessity or realization. Fortunately, this is the subject matter for urban planners who ought to prepare policy guidelines for the expansion of older cities and the creation of new settlements for the future.

The exponential growth recorded by the national vehicle fleet has increased the pressure on the demand for more space on the streets, that is, space for increased traffic, transit traffic, stationery vehicles and, occasionally, for vehicles no longer in service. Vital space has also been taken up by businessmen who park vehicles imported for sale (some of these vehicles might never leave the parking lot). Small traders have imposed themselves on the sidewalks hawking all types of goods while crowding out pedestrians.

The demand for space should be regulated so as to provide a harmonious environment for the functioning of socio-economic activities. The movement of goods and people is central to the success of all economic activities; this explains why priority space should be allocated to facilitate the movement and circulation of vehicles.

7.1 Space for motor vehicle traffic

Motor vehicle travel can be classified into three groups based on the speed of traffic circulation:

(a) Long distance trips with higher desired speeds (major streets, transit streets etc.);

(b) Local trips with lower desired speeds (district streets, distribution streets);

(c) Circulation with very low desired speeds (town centre, shopping centres, etc.).

The City Street in Camaroon

In addition to speed concerns, city street space is required for motor vehicle parking. Almost all our streets were designed before the motor vehicle boom. However, space is needed to accommodate very short-term parking for pick-up and delivery; short- term parking for shopping and business trips; and long-term parking (all-day or overnight) for employees and residents.

Vehicle access to property has already been reviewed under functions of the street (Section 2). The pick-up and delivery services necessary for business operations also need space on the street. However, the circulation of these categories of vehicle must be regulated to avoid congestion. Delivery times must be outside peak traffic hours. Unfortunately, the current system is indifferent to the traffic congestion created by delivery trucks circulating at peaks hours. This is inefficient and should be changed urgently because it imposes costs, discomfort and even safety hazards on other users. Garbage collection vehicles should be subjected to time regulation. For convenience, garbage collection should be done at night or during off-peak hours.

7.2 Space for public transportation vehicles

Urban development plans should provide sufficient land for public transport operations right-of-way. The availability of such land makes the layout of streets possible. In doing so, the minimum allowable distance from the centre line of the street to the beginning of any housing construction should be defined. This line on which all buildings on the street take their alignment is known as the "building line". It varies, depending on the location and function of the street. In some streets, it may be as wide as 15 meters (see details in the Annex).

Allocation must also be made for transport terminals. On existing streets, bus stops are designed and built to specification; there are technical reasons why bus stops are not marked out at random. The dynamics of slowing down a moving bus induce stress forces as it brakes and comes to a standstill. Take off after loading also requires forces to dislodge the bus from its stationary position. These forces are capable of deforming the pavement (especially at locations not initially designed to withstand them). This is visible on streets where artificial bus stops have been marked out by council workers.

There is a need for well-established, planned *bus routes*. The perennial lack of developed urban plans approved ahead of population settlement means that bus routes are absent. In populated areas such as Douala and Yaoundé, bus routes should be planned to give rapid and convenient access to as many parts of the town as practicable.[8] These bus routes should be suitable in width, alignment, and construction, and should include bus bays (stops). Therefore, in planning future development that incorporates bus routes, early consideration should be given to the likely demand for bus services.

Consideration and planning should also be given to *bus stops*. Bus transportation is relatively new in our urban settings. Its existence has been further complicated by the composition of the traffic stream, including taxis, trucks, truck pushers, private cars, motorcycles, and pedestrians.

As a matter of principle, bus stops should not be sited where their use might unreasonably interfere with the normal flow of traffic or restrict visibility at bends or junctions. Technically,

---

[8] *Roads in Urban Areas*, p. 47. Department of Environment. Scottish Development Department. The Welsh Office.

a bus stop on the approach of an intersection should be situated far enough away to ensure the following safety measures (cf. roads in urban areas):

(a) A waiting bus does not obstruct visibility rightwards from the main road to the side road, or the left from the side road to the main road;

(b) Traffic wishing to turn right is not obstructed by the bus (where buses turn right at the junction, it might be possible to incorporate a bus stop at the beginning of an additional lane for new streets or during major rehabilitation works);

(c) A bus requiring to turn left after leaving the stop has ample room to cross to the lane for left-turning traffic;

(d) Waiting buses do not interfere with the efficient functioning of traffic signals or the movement of traffic at a roundabout.

Standard practice requires that bus stops on opposite sides of single two-way carriageways should be staggered, preferably so that buses stop tail-to-tail and move off away from each other. These staggered stops should be 60 to 90 metres apart. Bus stops should be spaced at intervals of not less than 400 metres.

Bus stations and terminals must be planned. In Cameroon, the complete absence of any policy guidelines on the location of bus stations and terminals has played into the hands of bus owners. Their unwillingness to spend money on developing adequate terminals has introduced traffic chaos and unwarranted congestion in major towns. They install their terminals at street corners and junctions as well as load and offload without due consideration for other road users. No data are available concerning what it would take for competent authorities to engage in positive planning of urban areas. Research and planning would put an end to the current and constant shifting of responsibility between private operators and government authorities in Douala, Yaoundé, Bafoussam and Bamenda.

Emphasis must be given to the fact that many services generally radiate from the town centre. Because buses are required to serve both local and long-distance inter-urban traffic, their locations demand adequate study and research. Consultations with service providers and local authorities must be at the centre of all research and planning.

Technically, and sometimes for practical purposes, bus stations should be ideally located within or close to the central area, and they should have easy access to the distributor road network (this can be applied to future urban plans). Buses should be able to enter and leave the stations without delaying or endangering other traffic, preferably without having to cross or turn left against opposing traffic streams. The present locations of bus stations on curbs and off the sidewalks are untenable. The haphazard installation of bus pick up and drop off points has led to the concentration of bus traffic in small locations thus creating congestion of the nearby streets in Douala, Yaoundé, Bafoussam and Bamenda (see Photos 26, 30).

In order to make inter-urban bus travel attractive, bus operators should seriously consider the creation of all-day car parks adjoining their newly designed bus terminals.

The City Street in Cameroon

This would offer commuters an attractive alternative to the use of their private cars for long journeys. The Garanti Express bus terminal at Nsam in Yaoundé offers private car parking for individuals using their inter-urban bus services. In Bamenda, Amour Mezam bus terminal in Nkwen is a good example of a planned bus terminal that meets technical requirements. Consequently, this arrangement reduces the number of vehicles on inter-urban highways, thereby reducing the number of road accidents and senseless deaths.

Photo 9. Every space is good for buying and selling on the street.

Photo 10. New underpass and roundabout have improved traffic flow near the Governor's Office, Yaoundé.

Photo 11. With pedestrians strolling on streets, sufficient sidewalks should be provided to avoid accidents.

Chapter 7

Photo 12. Improved street junction complete with pedestrian crossing and traffic lights, Mvolye, Yaoundé, 2010.

Photo 13. The economic crisis of the 1980s left mountains of uncollected garbage on city streets.

Photo 14. Trailers parked on sidewalks abuse the street system.

The City Street in Camaroon

Photo 15. Garbage piles on city streets are accident and health hazards.

Photo 16. Disorderly parking in Bafoussam inter-urban transport system.

7.3 Space for pedestrians

Studies conducted in other countries indicate that in urban centres most of the traffic is by pedestrians. Their demand for space therefore requires an exhaustive examination. Foot traffic ranges from children in kindergarten who play as they move along the street to and from school, to octogenarians who sometimes have to be assisted to make it across the street. Children often have little or no sense of the danger involved in playing on the street, while octogenarians often have reduced motor responses.

Junctions pose the greatest difficulty in reconciling the varying interests of pedestrians with those of other road users. The complex movement of vehicles at intersections and junctions requires the vigilant attention of all motorists (riders and drivers) so that the lives of pedestrians are not at risk. These risks increase at unlighted junctions.

In the interest of safety, the design of junctions should incorporate guard rails, refuges and pedestrian crossings. By channelling or prohibition of certain turns (especially left turns), traffic movements at junctions can be simplified. Controlled signals at junctions have proven to be the most effective tools (for example, the post office roundabout in Yaoundé, Photo 18).

The segregation of pedestrians and vehicular traffic at points of heavy concentration will go a long way to eliminate or reduce any dangers of conflict occurring. This must, however, be accompanied by a strong dose of public education in schools and through the media (radio, television and newspapers).

7.4 Priority junctions

Whenever a line of movement crosses another, there is a point of conflict. Junctions are the intersections of these lines of movement and therefore constitute conflict points by their very nature. Controversy quickly arises as to who should cross first. In other words, who has priority at a street or road junction? In Canada, all road users have been educated to observe the priority of first come, first served, but there is the proviso that everyone must stop, look right, look left and look forward before allowing the first arrival to cross. Stopping at all junctions has become second nature to all road users in Canada. Road users in Cameroon must embrace and inculcate this behaviour.

Junctions should be sign posted. However, this is not always the case. It is generally accepted that when using a priority junction, traffic from the minor road is required to give way to traffic on the major road. This is controlled by the sign posting, "Give Way". At certain minor junctions, control may be achieved by the sign posting "Stop" and a horizontal white continuous line on the restricted lane. One wonders if our motorists learn anything in the driving schools, for 95% of motorists do not stop at the traffic signs posted at street junctions or at entry into roundabouts. This research study observed that almost all road signs restricting traffic or parking are almost always ignored. Doubting Thomases should visit the no-stopping signs in front of CAMTEL opposite the Central Treasury in Yaoundé, where traffic exiting from the roundabout is routinely blocked by taxi drivers who royally ignore traffic signs when picking up passengers. Another danger is at Carrefour EMIA where taxis ignore the place allocated to them, preferring to obstruct traffic flow at the junction. At the junction of street 1030 linking the Yaoundé City Council to the Hippodrome in front of the MAETUR building, one can see how those with restricted passes are forcing traffic to halt on a major street.

Priority junctions should have full visibility to the right and to the left between points one metre above road level. This would improve safety and avoid accidents. All fences and perimeter walls constructed around inner city properties should be built in conformity with instructions from the traffic engineer; fences must not obstruct visibility for road users turning from one street into another, or continuing by turning. Some examples can be seen in the evolution of priority junctions in Douala.

Diagram 4a shows a normal two-lane (i.e. two-way) street with no particular features to attract attention. The existence of this unknown and unnamed street on the outskirts of Douala which leads from Cité des Palmiers to the Administrative District and Business Centre in Bonanjo was anything but normal. Then in the early 1980s, the Government decided to build a new highway to link Douala (the business and economic capital) to Yaoundé (the political capital). The very high cost of expropriation for the right-of-way within the city forced planners and engineers to settle for the swamp lands on the fringe of the city as the starting point of this road project.

Contract No. 1062/AO/82-83 awarded to the consortium Razel Cameroon/Razel Frères and Boskalis Westminster was for the construction of Lot 1, the exit of the Douala-Yaoundé Highway. By 13 August 1985 when the construction work was partially handed over, the above street had been transformed into a T junction (see Diagram 4b, showing the intersection of the highway and the existing street). Where there are no traffic lights, there would be two possible points of conflict, at A and B (Diagram 4b).

7.5 Evolution of street junctions in Douala

Diagram 3: Types of street junctions

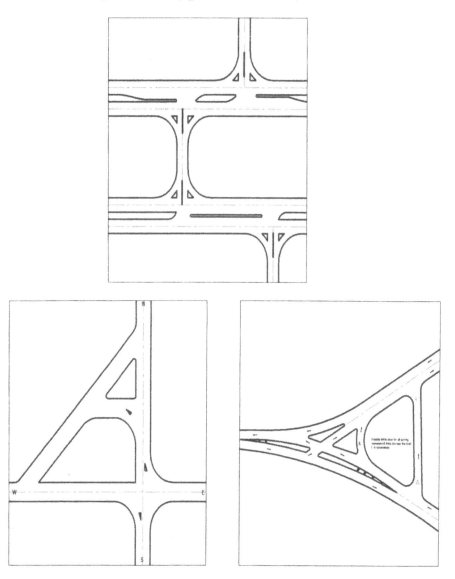

Chapter 7

Diagram 4a: Two-way streets

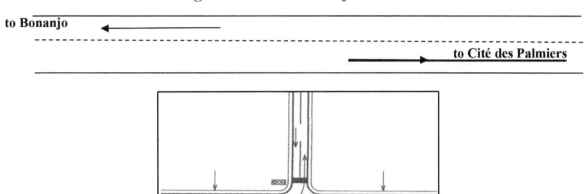

Diagram 4b: T-Junction   Diagram 4c: Dual carriage way

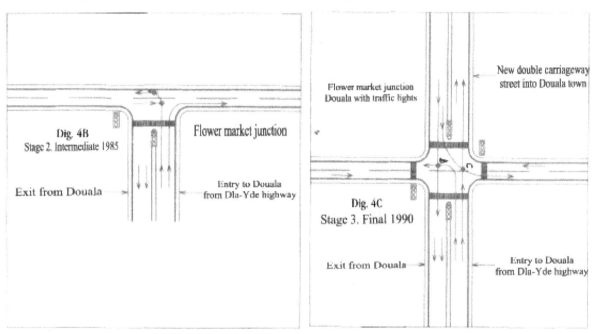

39

In stage three of the evolution of this street junction, after traffic lights were installed but wrongly programmed, there was a ghastly accident at this junction. The present author was commissioned by the Government to carry out a technical investigation into this accident. It was discovered that the green light was given at the same time to opposing traffic streams and this gave rise to the accident which occurred at C or D because the drivers coming from opposite directions were unaware that the others had been given the same green light at the same time (Diagram 4c). After long discussions, it was possible to convince the company that had installed the traffic lights to reprogram the sequencing so that green is given only to one directional flow at a time. This solution proved effective because no other accident has been reported at this junction.

7.6 Channelling islands

The *Highway Engineering Handbook* describes channelling and gives reasons why it is needed. Channelling separates two or more traffic streams and confines each stream to a definite channel. Channelling should reduce areas of traffic friction, separate points of conflict and correct irregular traffic movements by providing natural travel paths and permitting continuous safe traffic flow with minimum stoppage.

"The highest capacity and safest operations occur when traffic streams are parallel and move in the same direction. Capacity and safety decrease as the movement of opposing vehicles increases" (*Highway Engineering Handbook,* 1st edition, pp. 22–36). The objectives of channelling islands where necessary at priority and other types of junctions are to:

(a) Separate conflicting traffic streams;
(b) Assist traffic streams to intersect or merge at suitable angles;
(c) Control vehicle speeds (especially in crowded streets);
(d) Provide shelter for vehicles waiting to carry out certain manoeuvres such as turning left;
(e) Encourage drivers to take and stay in the correct path and deter them from straying into incorrect paths (undisciplined drivers are legendary for doing the wrong thing);
(f) Assist pedestrians to cross (while avoiding accidents);
(g) Reduce excessive land used for carriageways (because urban land is scarce).

The rehabilitation work on a major street at Mvolye in Yaoundé introduced channelling as an effective tool of traffic control. This measure halted the traffic chaos which had become a regular characteristic of the street.

CHAPTER 8
Use of traffic signals and signs

Uncontrolled traffic growth in urban areas has quickly caused priority junctions to become overloaded as a result of congestion. It therefore became necessary to install traffic signals to improve safety and increase flow capacity (even though their installation may actually lead to delays in the event of light traffic). Mention should always be made to the effect that our present street system was designed at the time when the purchasing power of the population was low, and also, because then the importation of second-hand vehicles was not allowed. (See Table 10 for data on average daily traffic.)

Table 10: Average daily traffic (ADT) on inter-urban paved roads in Camaroon, 2010 and 2011

Road section	Road kms	Road type	ADT*			
			2010		2011	
			April	Dec	April	June
Yaoundé to Obala	27	N101		4,166	5,104	
Yaoundé to Nkolefamba	20	N010		2,430	2,259	
Obala to Emana	26	N004		3,769	2,076	
Douala to Edéa	62	N3		5,833	5,114	5,066
Edéa to Pouma	47	N3		4,097	3,594	3,606
Douala to Bekoko	13	N5		11,859	11,489	9,644
Tiko to Bekoko	31			5,911	5,650	4,597
Loum to Bekoko	82			6,687	6,450	5,624
Bafoussam to Bamougoum	20			4,315	4,006	
Bafoussam to Foumbot	27	N006		2,069	1,914	
Mbouda to Santa	15	N?		2,947	2,527	
Nkongsamba to Melong	20			3,766	3,257	3,104
Limbe to Idenau	44	N?		4,549		
Mutengene to Limbe	14			7,594		

Source: Ministry of Public Works, Camaroon. *Traffic totals are only for to-and-fro journeys.

The standard sequence of traffic light signals has been determined as (i) red-red/amber, shown together, (ii) green, (iii) amber. The amber period is standardized at three seconds and the red/amber at two seconds. The two-second red/amber is provided only by the latest controllers; it is the older type that gives a three-second red/amber.

8.1 Vehicle-actuated signals

The latest types of vehicle-actuated signals include the following (cf. *Roads in Urban Areas*):

(a) *Minimum green period*. This is the shortest period of right-of-way which is given to any phase and is long enough for vehicles waiting between detectors and the stop line to get into motion and clear the stop line. The period is variable and depends on the number of vehicles waiting at the start of the green period.

(b) *Vehicle extension period*. The minimum green period may be extended by vehicles which cross the detector during the green period. Each vehicle, as it crosses the detector, extends the green period by an amount called the vehicle-extension period. This period depends on the speed of each vehicle as measured at the detector and is automatically varied to enable the vehicle to reach 3-6 meters beyond the stop line.

(c) *Maximum period*. A maximum period is timed off to limit delay on the waiting phase when there is a continuous stream of traffic on the running phase (after which the signal changes right-of-way independently). Maximum period is timed from the beginning of the green period if vehicles are waiting on other approaches or from the moment the first vehicle passes over the detector on one of the approaches. When necessary, the maximum period can be made variable so that the normal maximum can be extended, if at the expiration of the period, the traffic flow is above a pre-set value.

(d) *Inter-green period*. The standard inter-green period is 4 seconds, but when a longer clearance is necessary for safety, e.g., to protect clearing traffic, a variable clearance period can be provided. Additional clearance is then given while vehicles are clearing the junction but is omitted if there is no traffic clearing.

(e) *Early cut-off*. To facilitate a heavy left-turn movement from one approach, the green time of the opposing stream can be cut off a few seconds early. The duration of the early cut-off period can readily be adjusted by detectors of the turning traffic.

(f) *Late start*. An alternative way of facilitating a heavy left turn is to delay the movement of the opposing traffic stream for some seconds (cf. *Roads in Urban Areas*, p. 65).

8.2 Pedestrian signals

At street intersections, pedestrians come face-to-face with the problem of how to get across to the other side of the street. Given the attitude of most road users, pedestrians often find themselves stranded in one location for lengthy periods of time because no one wants to stop to allow them to cross.

At locations where intersections are controlled by signals, the green signal to pedestrians is displayed for a pre-set period of 6-10 seconds according to pedestrian flow. This is normally preceded by an all-red period of about 2 seconds and followed by an all-red period of 2 to 8 seconds, depending on the width of the street.

The sequencing of traffic lights at the Central Post Office roundabout (in Yaoundé) did not take into consideration the width of the street on 20th May Boulevard. Pedestrians often get cut off in the middle of the street when the pedestrian signal switches to red from green. Another problem is that traffic turning right into this street coincides with pedestrians crossing on a green light, causing untold confusion. A majority of drivers do not know that once the pedestrian signal goes green, drivers must wait for all those who have engaged in the zebra crossing to finish crossing.

Those who arrive daily from the villages or from other towns where there are no traffic signal lights become confused and stranded at zebra crossings. They do not know when to cross or which signal light (red or green) to obey. The majority of drivers do not know that at zebra crossings, once pedestrians have engaged in crossing, they must be allowed to finish. Instead, drivers tend to force pedestrians out of the zebra crossings. It is important to remember that pedestrians have the right-of-way on zebra crossings, and drivers must respect that rule.

Where there are no control signal lights at zebra crossings, drivers have the obligation to wait so that pedestrians can cross safely (especially children and the elderly). All zebra crossings must be signposted vertically as well as horizontally. Traffic wardens should be trained and stationed at zebra crossings near schools to assist children in crossing the street.

Vehicles should not be parked in the vicinity of intersections because they seriously affect the capacity and efficient functioning of these junctions. Parked vehicles reduce carriageway width and driver visibility; they are known to have caused fatal accidents at junctions, and even on the tangents on the street.

8.3 Traffic signs

Traffic signs are very important and therefore demand special discussion. Signposts are meant to guide, direct and control the behaviour of motorists on streets and highways. Traffic signs are made according to specifications and should be placed in clearly visible locations. They should always be located so as to allow ample opportunity for any necessary action or manoeuvre. Their number should be kept to the minimum required for proper guidance and control of traffic. With the exception of waiting restrictions and signs, some other prescribed types (warnings, mandatory and prohibiting actions and advance direction signs) should be illuminated by direct lighting where they are located within 45 meters of a street lamp (cf. *Roads in Urban Areas*).

The public needs to be educated about street and road signals and signs. A three-way approach to this education can produce lasting and durable results. The first step is to identify trouble spots and then provide signposts to guide, warn or direct road users as to the upcoming situation. Once signposts are in place, a vast media campaign on television, radio, in schools, and public places should be carried out to educate the population about why the signs are there, and what purpose they serve in traffic safety. The advantages of adhering to the instructions on signs should be outlined. Education should also include a list of penalties for those who violate or ignore traffic rules and regulations. The last stage is enforcement. City councils and governments must train and put in place special teams to ensure that all traffic signs are respected at all times.

CHAPTER 9
Roundabouts

Why do we design and build roundabouts? How do they function? And how do they differ from regular streets? These are questions we will try to answer in an attempt to explain the correct use of the roundabout.

Sometimes the simple provision of signal lights at junctions can solve traffic problems. At other times, roundabouts provide a smoother flow of traffic. Roundabouts are favoured and constructed in the following circumstances:

(a) There is a high proportion of left-turning traffic at the junction. A four-leg junction roundabout may need less land than a traffic-signal junction if left-turning traffic exceeds about 30% of all approaching traffic;

(b) There are more than four approaches to the junction. Examples are the Central Post Office, Hilton Hotel, Ministry of Posts and Telecommunication, and Ministry of National Education roundabouts; the Nlongkak roundabout; the roundabout at the Mfoundi Divisional Office; the Bonabéri roundabout in Douala, the *Carrefour des Deux Eglises* in Douala; and City Chemist roundabout in Bamenda; the BEAC roundabout in Bafoussam.

(c) The approach width is so restricted that it would be impossible to provide separate lanes for through and turning traffic.

(d) Other junctions are so near that there would be insufficient space for the formation of queues.

(e) Where there is a Y junction layout (it lends itself to roundabout design). At sites where control by roundabout or traffic signal is equally feasible, a roundabout may be a safer and less restrictive option.

Roundabouts function as traffic distributors for the streets converging onto them. Roundabouts are different from regular streets because traffic circulation goes around

The City Street in Camaroon

circles in many lanes. They also have many exits and entrances. There is a central island that is round in shape and acts as a speed regulator.

Roundabouts have the tendency to lock when overloaded or when misused by vehicles that position themselves across the lanes. Therefore, it is recommended that at the design stage, roundabouts should be allocated adequate reserve capacity to meet future peak flows; attempts should also be made to reduce the possibilities of locking. The only experimental method that has given promising results is the placement of signs requiring approaching traffic to give way to traffic already in the roundabout. Since road users often do not read road signs or deliberately ignore the instructions on them, an additional measure should be the creation of speed brakes at the entry into those roundabouts not controlled by traffic signal lights. The speed brakes could eventually be removed when drivers systematically respect the signs.

Another sign should be posted near the roundabout that reads: "You do not have right-of-way". Sign posting would ensure that locking is reduced or eliminated while maintaining capacity traffic flows, even in overloaded conditions. The phenomenon of locking appears frequently at the roundabout in front of the Ministry of Posts and Telecommunication and the Ministry of National Education when all traffic grinds to a halt. All roundabouts prone to locking in peak times should have traffic signals on the approaches. The Bonabéri roundabout in Douala is another example where locking occurs frequently.

A serious traffic safety problem exists at a roundabout in Yaoundé. In an open violation of traffic regulations, a traffic light was installed facing the roundabout located near the National Security Head Office; the light was meant to regulate exit traffic. Unfortunately, when the light turns red, exiting traffic blocks circulation in the roundabout. The result is the locking of the roundabout bringing all traffic to a stop. This traffic light should be removed immediately (no traffic engineer would have authorized such an abnormality).

In some countries, education and the appropriate use of signs have reduced accident rates at roundabouts by more than 40%. There is also an urgent need to provide real traffic management education, for example, by conducting seminars or giving information through the media.

Street lighting columns and traffic signs on the central island and the periphery of a roundabout should be sited where there is little risk of their being hit by out-of-control vehicles. Their location at the centre of the raised island has proved ideal (cf. F.V. Webster and R.F. Newby. "Research into the Relative Merits of Roundabouts and Traffic-Signal-Controlled Intersections". Proceedings of the Institution of Civil Engineers, January 1964).

Because of their size and number of lanes, roundabouts often contain pedestrian crossings of the zebra type sited at their entrances and exits. Motorists and pedestrians should be educated on the use of roundabouts, and driving schools should emphasize them.

To avoid the absolute confusion and chaos that sometimes exists at roundabouts, the following measures need to be taken (see Diagram 5 for rules):

(a) In Cameroon, driving is on the right side of the street or highway. This principle must be strictly observed when in the roundabout. Exiting from the inner lane of the roundabout amounts to driving on the left (this is not allowed);

(b) Because the roundabout is not a street, motorists must be educated to understand that the rules that apply in the roundabout must be respected at all times. No crossing of lanes to exit is allowed;

(c) Drivers coming into the roundabout must give way to the traffic already circulating inside the roundabout because it is they who have right-of-way not those entering. This is what is known in the driving code as "priority to the left";

(d) All traffic exiting the roundabout must use the last outer lane on the right side to avoid crossing lanes and blocking the continuous flow of traffic (that is: exit in lane 4);

(e) It is strictly forbidden to drive across the lanes inside the roundabout in order to leave the roundabout without using the extreme outer lane reserved for that purpose and marked "exit". This last outer lane is known as the "Exit Lane" and must be used for that purpose at all times;

(f) Where there are three or four lanes, the inner lanes (that is, lanes 1 and 2) close to the island are used by traffic moving around, without the intention of exiting;

(g) To exit from these lanes, motorists must change lanes until they get to the outer lane, that is, the extreme lane on their right. This is what is meant by continuing to drive on the right in the roundabout.

The City Street in Camaroon

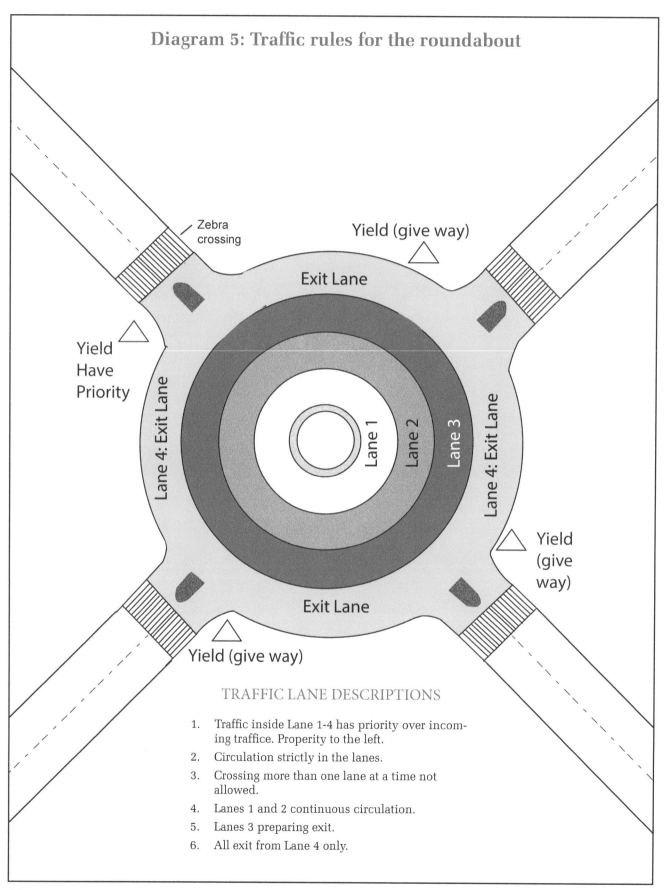

Chapter 9

Photo 17. Bonabéri Roundabout, Douala, with the white car in traffic.

Photo 18. Central Post Office Roundabout, Yaoundé; traffic signals and zebra crossings provided.

Photo 19: Street allows light and air into buildings at Oxford Street, London, November 2011.

Photo 20: Roundabout on Nkolbison Road, Yaoundé.

The City Street in Camaroon

Photo 21. Old street, Mvolye, Yaoundé

Photo 22. Old Avenue Kennedy, Yaoundé

Photo 23. Improved street, Mvolye, Yaoundé

Photo 24. New Avenue Kennedy, Yaoundé

Chapter 9

Photo 25. Diagonal crossing of lanes in roundabout obstructs traffic flow near Governor's Office, Yaoundé (see taxi dropping off passengers inside the roundabout)

Photo 26. Spontaneous creation of interurban bus terminals causes traffic congestion, Mvan, Yaoundé

Photo 27. Confused traffic flow at the roundabout due to diagonal lane crossing to exit near the Governor's Office, Nlongkak, Yaoundé

Photo 28. Floating mat of debris drifting to the ocean near the roundabout at Custom Headquarters, Douala

CHAPTER 10
Location of public utility services

10.1 Utilities

The accommodation of public utility services on the streets requires careful planning and the input of engineers, architects and town planners. The absence of this consultative body has created a no-man's-land in the use of urban space with the attendant disorder that prevails almost everywhere now. Lack of knowledge about the location of these lines means that pipes are destroyed every time workers dig to install new cables.

There is no doubt that the existing streets and their design have outlived their usefulness. The existing method whereby service lines (electricity, water, telephones, and TV cables) are placed indiscriminately has led to constant digging for repairs, obstruction of traffic, premature destruction of the streets and pavements, etc. The positioning of underground mains should henceforth be subject to regulation. The depths at which they are buried under sidewalks, their spacing and accurate recording of their locations should facilitate maintenance operations and free carriageways from untimely destruction. Location of mains should be under the sidewalks; locations should be shown on the "as built drawings" produced at the end of construction or rehabilitation of a street. The construction of new streets or the rehabilitation of existing ones should afford the needed opportunity for a transition. The placing of distribution mains in duplicate, one on each side of the pavement, will diminish the need for lengthy service connections under the carriageway.

However, where it is necessary for service connections to cross the road, they should be laid before the carriageway is constructed. On the recommendation of civil engineers, mains should normally be located in the following order between the street boundary and the kerb: electricity, gas, water, telecommunication. This placement ensures reasonable working space for each utility and accommodates junction boxes and hydrants. It should be emphasized that electricity and telecommunication mains will require additional

The City Street in Camaroon

cover if placed beneath the carriageway (cf. Report of the Joint Committee on Location of Underground Services. Institute of Civil Engineers, London, 1963) (see Diagram 6). The depths should be in accordance with the specifications on street design. In order to facilitate maintenance operations, the location of utility service lines should be clearly marked on the contractor's as-built drawings. Copies of these drawings should be addressed directly to the owners of the utility services.

Diagram 6: Location of public utility service lines

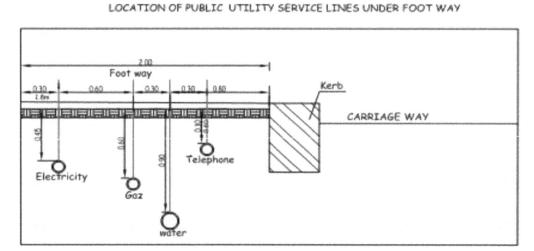

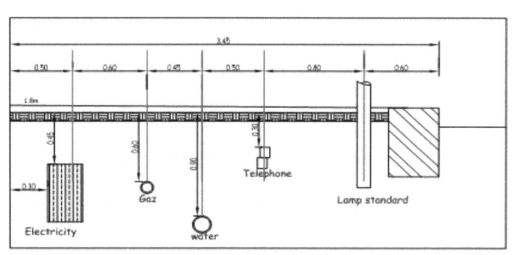

10.2 Sewers

Technical norms require that sewers should preferably be routed along streets of little traffic importance and through open spaces. Sewers should be laid in straight lines between manholes. Sufficient reservation should be made in the clearance to accommodate future services. They should be deep enough to ensure that branch connections can be made without interference with other services.

The current unhealthy practice whereby used water from households is discharged directly onto the streets should be urgently abolished. This degrades the surrounding environment, produces an enabling breeding ground for mosquitoes in neighbourhoods, and results in unpleasant odours.

10.3 Surface water drainage

Existing street designs were modelled completely on European norms. Even as we write, nothing has been standardized. In the tropics, there is need to study the surface water run-offs produced by torrential rainfall. During peak rainfall, weep holes are ineffective because they cannot ensure the rapid discharge of surface water without causing discomfort or danger to street users. When there is considerable surface run-off, water collection by means of gullies placed behind the kerbs with side-entrances has proved ineffective. The overflows have caused untold destruction to streets, especially those on slopes and in low-lying areas. Adjacent property has often been affected by flooding as well.

The Mfoundi River begins in the form of small brooks. Founded around seven hills, the City of Yaoundé appears not to have major drainage problems until these brooks or collector streams hit the low-lying areas and form the Mfoundi River that flows through the city centre. Part of the water flows into a closed up concrete canal below 20th May Avenue, while the other major tributary flows down along the Central Railway Station to the Mfoundi Market and on to the Central Post Office roundabout. From here it becomes a stream running through marsh lands for some kilometres before leaving the city.

In the low-lying areas, water stagnates as refuse and reeds intermingle to slow down the flow of the river (PIC 32). This creates pools of standing water everywhere. These pools become the breeding ground for millions and millions of mosquitoes that spread malaria and other diseases throughout the City. Fortunately, the government delegate to the Yaoundé City Council proposed the construction of a concrete channelling canal for the river from its exit behind the Central Telephone Exchange in the centre of town to the Old Airport. This project created a concrete riverbed and stone masonry walls that ensure continuous flow of water. The canalization of the Mfoundi River destroyed the habitat of billions of mosquitoes and freed thousands of inhabitants from the ravages of malaria (see PIC 32). Additional attention should be given to the drainage structures in the coastal areas of Douala, Victoria, and Kribi where zero grades necessitate special drainage to cope with heavy rain. This would reduce the risk of flooding caused by backloads of water.

CHAPTER 11
Commerce along city streets

11.1 Demand for commercial space

The demand for commercial space on city streets is insatiable because the numbers of those asking is steadily on the rise. The crisis of unemployment is driving more and more people onto the street where they can set up shops or simply become street vendors. Many of them are university graduates or holders of other certificates; many cannot even afford such enterprise.

There seems to be no law on the street where its invasion has taken an undefined form. There are people hawking their goods on mobile trucks, some selling phone cards and phone calls, some preparing and selling food, some buying and eating on the spot, others repairing all kinds of devices, and crooks trying their luck at cheating passers-by. They all scramble for space on the street where they can settle in order to eke out a living by laboriously practising some trade.

A recent attempt by the Yaoundé City Council to expand the central market (Marché Central), almost brought about a dog-fight. The number of those requesting shops outstripped the would-be available shops by a ratio of 200:1. The decision was therefore taken to draw lots so that the lucky winners could rent the shops. The constraint was that those who won the lottery had to pay upfront one year's rents before construction even started.

A new breed of street people has surfaced recently: the commercial motorcyclists commonly referred to as *bendskins*. Their demand for space ranges from the possible to the absurd. They have made the street their personal property. They ride on the sidewalks and separation islands, violate traffic rules and stop lights, and overtake on all sides. They are arrogant and aggressive. These street users have never been schooled in the spirit of sharing space, nor have they been informed of traffic regulations. They actually refused to be trained to observe traffic rules; they have just imposed themselves and are setting their own rules. In short, they have simply invaded the streets like locusts attacking a farm and are in the process of destroying our social fabric. The question about what amount of space can be allocated to motorcyclists and how that space should be managed is crucial. Meanwhile, their numbers are escalating every day, and the death toll from motorcycle-related accidents keeps mounting.

The availability of space on the street is limited. Consequently, the authorities should make maximum use of the space that can be found by regulating demand. Because of the limitation on both the land available for the street and the size of the street itself, a detailed study of the situation is imperative. Only a careful study that brings together all the interested parties (those demanding space and those who can provide it) can bring about a rational response. Town planning should be at the centre of all new developments in cities.

11.2 Filling stations and garages

Fuel stations are a prominent feature on the city landscape. They are the storage locations were vehicles are refuelled. Some fuel stations also offer regular vehicle servicing within or outside the cities. Their design, location and construction must be subjected to the special requirement of fire-proofing. They must be located away from residential buildings. Their operation must conform to the norms of environmental safety standards. The disposal of toxic waste such as used engine oil and spent oil and fuel filters must be regulated and controlled to prevent pollution and any health hazards. The provision of air pumps, toilets and stand-by water taps should be compulsory and their continuous functioning policed by inspectors of service.

Garages are to vehicles what hospitals are to humans (a place for maintenance and treatment). Broken down and ailing vehicles must be towed to a garage for repairs or maintenance; they should not be abandoned on the street. The current practice of setting up garages anywhere along the street, on the sidewalks, or worse still, repairing vehicles on the street itself is untenable and should be abolished. The town councils should provide a model design for the construction of low-cost garages. The size of a garage will depend on the expected number of vehicles to be handled at a given time.

The location and operation of garages should be regulated to control the disposal of used parts and fluids (oils, fuel, and other substances) discharged during maintenance. It is imperative to ensure that the functioning of garages conforms to environmental safety regulations at all times. References and requirements here are to mechanical garages as opposed to ramp garages.

11.3 Billboards

A substantial amount of space on the city streets, on buildings, and even on electric poles is allocated to billboards. They are large outdoor boards for advertisements of all kinds. Some of these advertisements are written or ingrained on large cloth banners and hung across the street on poles (existing or improvised for this purpose). They are very often forgotten and left in place long after the event on the advertisement has passed.

These billboards carry concise notices advertising goods or offering services. The lack of regulation means that the boards are often placed in locations where they obstruct the vision of motorists circulating on the street. Billboards appear almost everywhere. On sidewalks, they very often force pedestrians to divert their route or bend down to pass under them. Others are posted on the islands of dual carriageways; depending on their size, these might seriously obstruct the vision of motorists.

The important questions as to who determines billboard size, location of posting and the duration of their stay currently have no answers, but these issues need to be resolved soon. In some cities elsewhere in the world, revenue collected from street billboards offsets the cost of street lighting. For the purpose of efficient management and appropriate standardization, the entire aspect of allocation of space for billboards should be privatized. Detailed studies should be conducted to establish how many billboards actually exist in each city. From this data it would be possible to estimate the possible revenue that could be generated from adequate management. Tenders could then be solicited from interested parties.

Directional billboards are necessary to guide and channel traffic through and around cities. Currently such billboards are conspicuously absent. Every attempt should be made to ensure that they are signposted at the required locations.

CHAPTER 12
Naming streets

How do we direct visitors to where we live? When we shop and our goods need to be delivered, what is the forwarding address? On what city street do we live and where in town is it located? What is our house number and the city street name? These are questions we must now address. Giving a name to a street is similar to naming our offspring. A street name is like a trademark by which goods and purchases are identified. Street names refer to one and only one particular location in a given city or town. The street name must therefore be seen as a landmark or an identity tag placed at a particular location as a directional sign to guide people wishing to arrive at that destination. Street names should be clearly visible at all times, day or night.

Once given, street names must be embedded as reference points on fixed benchmarks. These names are unique and specific to the towns where they belong. Apart from their importance as sign posts, the construction of street signs can and do offer employment; the naming of streets followed by the numbering of houses can lead to the establishment of mail delivery services at home as well. The existence of street signs facilitates the provision of ambulance services or fire-fighting equipment. Additional security of the population would be guaranteed by the actual numbering of buildings on the streets.

What names are chosen for streets? The answer is simple. Any name that enables the citizens to identify the street should be welcome by all. Research into the names of streets can bring out an amazing catalogue of the socio-cultural, historical and even commercial events that influence the naming of streets. One example is the famous shopping street in the heart of the City of London. Oxford Street, known as one of the most popular business attractions in Britain and Europe, is home to numerous shops visited by millions of tourists and local citizens every year. It is interesting to note that in the twelfth century, this now famous street was a notorious road taken by prisoners to Newgate Prison. In the eighth century, the name became established as Oxford Street based purely on a historical coincidence, that the land on the northern side had been acquired in 1713 by Edward Harley, 2nd Earl of Oxford (Wikipedia, retrieved 18 February 2012).

In Cameroon, not just streets but entire districts have been named after historical figures. Examples are Akwa in Douala named after the famous King Akwa who signed the German-Douala Treaty of July 1884 when the first Germans arrived in Cameroon. By the same token, Boulevard Ahmadou Ahidjo, a dual carriageway in Douala, was dedicated to the first president of the Republic.

In Yaoundé, an entire zone was named after the commercial activity that was established there. Bastos, the opulent neighbourhood was so named for the cigarette manufacturing company that was situated there before independence. The American Tobacco Company has since liquidated and left Cameroon, but the name may remain forever as a reminder for future generations about what used to be there. Quartier Fouda was named after the industrious first mayor, Mr Andre Fouda, whose historic input into the development of Yaoundé is widely known and recognized.

In Bamenda, Commercial Avenue was so named because of the activities carried out around there. By contrast, Foncha Street, still in Bamenda, was named after the man who created it and who also happened to have been the Prime Minister of Southern Cameroons, and later Vice-President of the Federal Republic of Cameroon. Cow Street is another example of a street acquiring the name of the predominant activity on it. It was famous as the site where cattle breeders from all over Bamenda Province brought their cows to sell at the cow market, and people came from the neighbouring villages to buy them.

Kumba in the South-West Province, is one of the best examples of a cosmopolitan town. This has been reflected in the names of its quarters and streets. Hausa Quarter, Ibo Quarter, Bafaw Quarter, etc. are all named and devoted to those who have inhabited these areas. Commercial activities have sometimes influenced street names as has already been indicated, Alaska Street in Kumba originated from a form of icy sweet or candy (an Alaska) shaped in the form of a cylinder. It was concocted from a mixture of sugar and canned fruits, then frozen and later sold. The street took its name from the product (Alaska). Cultural identities have also shaped street naming. The recognition of Mbo Street in Kumba stems from the historical fact that people of Mbo origin were the first to settle and build houses in that location. There is also the creation of SOBA Street named after the Sasse Old Boys Association because the first occupants of the street (SOBANS) used to hold their ex-student association meetings there (to promote their *alma mater*).

While the naming of streets in some towns has proceeded smoothly; in others, it has been marred by unqualified suspicion about whose name should be used and why? This has created a virtual stalemate by default in city halls. However, city authorities in Bafoussam (West Province) have adopted the serial numbering system similar to that of Tiko in the South-West Region. In Nylon Quarter which begins on the dual carriageway linking Bafoussam to Bamenda (just behind Continental Hotel), the streets are known as 1st Street, 2nd Street.... 10th Street and so on. The houses on them are also numbered.

Because these streets begin or end on the dual carriageway, it has been easy to remember them. This successful experiment should be continued and extended to other districts. The system could also extend to cover the whole town. Names could be assigned to roundabouts and street junctions. For instance, roundabouts could be named according to divisional headquarters: Mifi, Ndé, Noun, Menoua, etc. Major road junctions could acquire their names from rivers, trees, mountains, administrative sub-divisions, etc.

Faced with inordinate delays in the production of traffic road maps, an innovative solution would be pleasing to all. By giving the names of local trees, mountains and rivers to our streets, we are actually on neutral territory. The people have, in desperation, begun to give whatever name they can to identify various locations in towns and cities. For example, one might hear names of streets or other landmarks like Tonton Bar, Transformateur, *Carrefour des Deux Eglises* (referring to a street junction that has two churches), or *Carrefour des Trois Morts* (immortalizing a junction where three people died in an accident). These unconventional names have become known to the population and act as reference points in the same way that conventionally assigned names do.

When comparing this to an exercise that allocated numbers randomly to some streets in Yaoundé, it is easy to explain why these numbered streets are not remembered. The reasons are simple: without an anchored reference point, people do not remember how to locate a given place. How for instance, does one relate Street 1085 to Street 880, both located in different parts of the city and unconnected? These numbered streets have no linkage between them and, therefore, exist in perpetual isolation serving no useful purpose for society. The exercise cost tax payers hundreds of millions of CFA francs but made no difference in their lives.

It is essential to have a framework which anchors the naming system and allows it to operate. The framework should identify and name street junctions, roundabouts, parks, and monuments (if any) as reference points or anchors. Diagram 7 illustrates such anchorage. Suggested neutral city street names have been identified and listed in the Annex.

The City Street in Camaroon

Diagram 7: Model street junctions and names

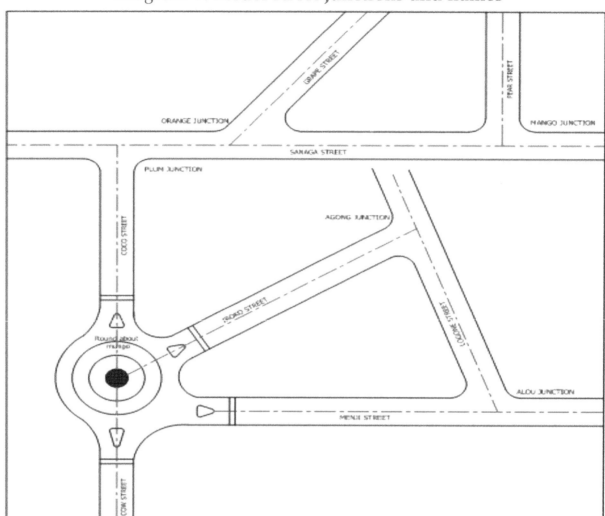

In London, there is an underground station known as Seven Sisters. This is an illustration of how names become labels to guide those navigating through cities. The origin of this unusual name begins in the Scottish Highlands and concerns the secret of oil. In August 1928, three men (a Dutchman, an American and an Englishman) had an appointment at Achnacarry Castle. The Dutchman was Henry Deterding, a man nicknamed "the Napoleon of Oil" and a founder of Royal Dutch Shell. The American was Walter C Teagle representing the Standard Oil Company. The Englishman, Sir John Cadman, was the director of the Anglo-Persian Oil Company soon to become BP. These three formed a great cartel[9] whose purpose was to dominate the world by controlling its oil. Four others joined them to form the consortium for Iran known as the "Seven Sisters" (cf. Business: the seven sisters still rule, *Time*, 11 September 1978). The London Transport Authority adopted the name and gave it to a subway station in honour of BP, which was a founding member. The lesson here is that names can and do have a history of their own, even though some are just identification labels.

---

[9] *Documentary. The secret of the seven sisters aired by Aljazeera TV, 25 April 2013. Retrieved 4 March 2013.*

It is therefore acceptable to have, Mango Street, Plum Street, Coco Street, Coffee Street, Yam Street, Orange Street, Iron Street, Love Street, Brothers Streets, Drivers Street, Student Street, John Street, Begue Street, Menji Street, Legwe Street, Alou Street, Njeh Street, Ndungatet Street, Mankem, Iroko Street, A&B Street, Hospital Street, Nurses Street, Engineers Street, AJ Street, Corn Street, Cotton Street, Lemon Street, Eru Street, Pineapple Street, Zebra Street, Elephant Street, Monkey Street, Lion Street, Dog Street, Cat Street, Bird Street, Forest Street, Lake Street, River Street, Nkong Street, Cow Street, People Street, Eagle Street, Tiger Street, Church Street, AJM Street, Lekong Street, etc.

CHAPTER 13
Pollution and congestion

Pollution and congestion of city streets have attained their own notoriety as negative forces in the Cameroonian society. Both phenomena are related to and a result of massive population movements. Highway traffic jams, queues, population crowding as well as air, water and noise pollution are all forms of congestion resulting from the common consumption of public goods such as streets and highways.

During traffic congestion, there is the experience of deterioration in quality of life due to long commute times, reduced safety, and increased mental and physical tension. A simple automobile trip to and from home becomes complicated due to journey time, conditions of movement and the presence of unexpectedly numerous users. The disruptive effect to traffic flow of delivery trucks, taxis not in queue and motorcycles is another contributing factor. Such traffic must be harnessed and brought under control.

The vehicle fleet in Cameroon includes vehicles aged 25 years and above; their damage to the environment has become a major concern. During traffic jams, traffic virtually comes to a standstill (in Douala and Yaoundé, this may last for hours); vehicles have to idle all this time. Engine idling is inefficient and contributes to the problem of air pollution.[10] An additional warning concerns the effect of diesel exhausts which have been linked to lung cancer, chronic bronchitis, aggravated asthma and other health problems. Diesel exhausts have also been identified as a greenhouse gas that is contributing to global warming.[11] The lesson from all this is simple: increased congestion, especially from old vehicles, is a health hazard for people. The policy of importing vehicles of all ages needs to be reviewed.

Pollution does not only concern air. The physical environment is aesthetically disfigured by mountains of litter. Uncontrolled levels of noise produced by random speakers and amplifiers spew all types of sounds at pitches unfriendly and unhealthy to the human ear. Some parts of towns produce odours which would send inhabitants fleeing to better habitats.

---

[10] Idle Reduction Environmental Protection Agency. Retrieved 06 March 2008.
[11] Health Effects of Diesel Exhaust Particulate Matter. California Protection Agency. Archived from the original 07 July 2007. Retrieved 14 April 2008.

It can be said that although the capacity of a public good may be natural or given by nature as a free good, deliberate human action can either increase that capacity or negate its quality. For example, the irrational and disproportionate misuse of streets by some motorists (especially taxi drivers and vendors) tends to defy the definition of a public good. That is, their excessive misuse deprives others of the use of this good, which should not be the case with public streets.

Either because of lack of education or purely out of insensitivity to the needs of others, some people use rivers as sewers for noxious household refuse. In extreme cases, they even misuse gutters designed to evacuate rain water run-off. They ignore the fact that other people downstream simply want to drink the water. During the rainy season, the rivers in towns become floating mats carrying tons of solid and liquid waste. In the process, some of the garbage is downloaded onto riverbanks or low-lying areas when the flow velocity decreases. Anyone standing near a riverbank in any of the major cities during heavy rainfall can witness the massive quantities of debris floating downstream (see Photo 28).

The forceful discharge of pollutants (especially smoke, hydrocarbons, plastics, etc.) imposes costs on the members of society. Because these costs are never adequately imputed to the source, the community is constantly being polluted. For example, the discharge of used engine oil and filters from filling stations and garages is not regulated. These stations are supposed to be equipped with water, air pumps for tyres and equipment for the disposal of waste products from the servicing points. Detailed research is needed to draw up adequate regulations to prevent abuse and the continuous imposition of pollution on society by commercial and industrial companies.

Industrialization has brought with it serious problems of air pollution in urban areas. Although this is affecting people now, the general impact may not be evident because of the absence of any research to identify any negative impact. Is there a relationship between levels of pollution and the increase in air-borne illnesses? Only research can provide answers.

About ten or fifteen years ago, traffic congestion was almost unheard of in cities in Cameroon. Traffic congestion has been compounded by the behaviour of road users. A liberal importation policy has accelerated the growth of the vehicle fleet by flooding the streets and highways with old vehicles. In most cases, the size of the street is adequate to handle existing traffic, but this is being disrupted by disorderly parking, lack of designated bus and taxi stops, incursion into the carriageway by kerbside parking, and the inadequacy of available space for off-street and on-street parking. These problems were not envisaged during the creation of streets.

By introducing kerbside parking on streets whose initial design did not incorporate that element, city councils and governments have inadvertently precipitated congestion in the streets. The whole exercise should be revised and carried out with the assistance of traffic engineers. It is inconceivable to have vehicles parked just before traffic lights or too close to street junctions. It now takes more time to drive across Douala (a short distance of about 10 to 12 km) than it takes to drive from Edéa to Douala (a distance of over 62 km).

Statistics indicate that the average daily traffic count entering and leaving Douala on the above highway is 5036 vehicles (2011 statistics). This urgently calls for the construction of a 3x3 divided carriageway and subsequent creation of a bypass (see Table 10).

Disregard for the driving code (for example, the obligation to drive on the right at all times) has also become a stumbling block especially in roundabouts. At street junctions and in roundabouts, a couple of road users who ignore the rules can bring all traffic to a standstill.

The topics of pollution and congestion are mentioned here because they impose costs on both individuals and society. Urban planners must examine their negative impact and seek solutions. Smoke is a typical example. Those who live near factories or production plants are constantly bombarded by smoke. Toxic smoke may cause illness to those who inhale it, but the cost of medical treatment is imposed on neighbours who have no say in what factories do or how they are managed. People are condemned to bear the negative impacts of industries just because they happen to be neighbours. The separation of industrial areas from residential areas must be reinforced in our urban planning policy.

It has been indicated that traffic engineering attacks the problems of traffic accidents and congestion from two approaches: constructive and restrictive. Here is a practical example from the planners and engineers of the City of Los Angeles, California. To solve the endemic problem of congestion and traffic accidents, the city authorities authorized the construction of additional lanes on the main street carriageways. One lane (the inner lane) in each direction was reserved for carpooling (that is for cars carrying more than one person). All other cars were forbidden from driving on that lane, even when the lanes were empty of traffic. Very heavy fines were imposed on those caught violating the law. The result was less congestion on the other lanes as more and more people were attracted to carpooling.

What can we learn from this rich experience? The restrictive approach can make the use of existing streets and highways more efficient, especially if traffic regulations are supplemented by traffic control devices. All we need to know is that we cannot re-invent the wheel, but we can learn from others about how to use the wheel effectively. The restricted approach could be used in regard to the policy of vehicle importation in Cameroon. Limiting the age of vehicles that can be imported to say 1 to 5 years would restrict the size of the vehicle fleet and reduce the number of ageing vehicles that cause the most pollution in the country.

CHAPTER 14
Future street design

The design of city streets will certainly be based on function and classification. Design must take into consideration all the elements of safety as well as appropriate degrees of traffic segregation to reduce the risk of conflict between the users, that is, motorists and pedestrians. Design must also concern location and alignment, traffic speeds, sight distances or vision, gradients, horizontal curves, etc. Where a street is on a curve, particular effort must be made to locate junctions away from steep gradients. It is obvious that the width and layout of a street or road will depend largely on the type, volume and speed of the traffic it will carry.

Carriageway capacity must be considered. The speed of traffic in towns is lower than that on rural roads, and there is less overtaking because of the restricted size of the lanes and the density of traffic. Civil engineering students and future urban planners should pay particular attention to the fact that there is a major difference between designing the main traffic routes for built-up areas and routes for rural areas.

As a matter of principle, urban roads should be designed to be safe and to allow the free flow of traffic at acceptable speeds. This calls for appropriate planning of the urban road network as a whole, including forecasting of future traffic volumes and an adequate concern for parking locations. The design of main traffic routes (streets) in built-up areas should be based on peak-hour demands. This is because the socio-economic and other activities of those who live in these areas generate traffic peaks at certain hours. Generally, in the mornings, people are rushing to work; dropping off children at school; going to the market, the bank, the hospital, or the hairdresser. The same rushing about occurs at the end of the day's activities when the urge to get back home quickly becomes the desire of everybody.

By contrast, the design of routes in rural areas is based on the average daily traffic (ADT) recorded during traffic census studies. It is rare for peaks to be produced on these routes at predetermined times. Future designs of city streets must integrate kerbside parking, bus stops, and possibly motorcycle lanes separate from the carriageway. The example of a motorcycle lane in Edéa is laudable, even though it has not been put to effective use because of poor enforcement and lack of education. Motorists use it now for parking their vehicles and obstructing motorcycle traffic.

14.1 Traffic census

A traffic census or survey includes field studies of traffic carried out at predetermined locations on existing streets and routes for a certain period of time. The statistics collected are used for design. The time of day and the period of the year should allow the recording of all possible traffic movements. For this reason, a traffic census should not be done in the rainy season or at night.

Traffic statistics are vital for the rehabilitation of city streets and the design of new ones. Their collection should therefore be organized at determined intervals so that trends in their growth can be easily interpreted. It is difficult to determine whether any traffic census has been conducted in any major city in Cameroon over the last three or four decades. However, the truth is that rational planning for the rehabilitation, creation or total expansion of city infrastructure must be based on traffic statistics. Without basic statistics, city planners turn to guesswork and improvisation. The result is that inadequate measures are used to resolve the myriad problems confronting city streets, including traffic congestion, rapid environmental degradation and urban traffic accidents.

14.2 Rehabilitation of a city street in Bonabéri, Douala

Inadequate piecemeal solutions or outright postponement of durable solutions to urban street problems will not make them go away. They only complicate or worsen the situation. In Douala, for instance, the rehabilitation of a city street inadvertently led to civil disturbance in Bonabéri. People had anticipated that the perennial traffic congestion that usually paralyzed transport movement in their part of town would at last be alleviated by the street rehabilitation. Unfortunately, this did not happen. Instead, the rehabilitated street turned out to be narrower than the old one. What follows below is a detailed explanation of what went wrong. The discussion also proposes a better, more successful scenario.

The engineering design of highways and city streets is based on well-established norms founded on the desire that the infrastructure to be put in place should meet community needs. This is especially so in cities. In his new book, *If Mayors Ruled the World*, Benjamin R Barber argues that "the city, always the human habitat of first resort, has in today's globalizing world once again become democracy's best hope". This document does not elaborate on the definition of the city or town because it is based on the city street as a structure within the city. Suffice it to say that it is necessary to ensure that the city continues to exist through the efficient management of city streets and the elimination of the chaos that has built up over the years in our cities. Some of this might require new technology. In Singapore, the city has developed an electronic road pricing (ERP) system that greets motorists as they drive into the city centre. An electronic billboard tells drivers every minute how much it will automatically charge them while driving

Chapter 14

down town. It constantly adjusts the toll based on the number of cars that can comfortably fit on the streets (Thomas L. Friedman, "Calling America", *International New York Times,* 4 November 2013, p. 9).

Diagram 8: Some street cross-sections

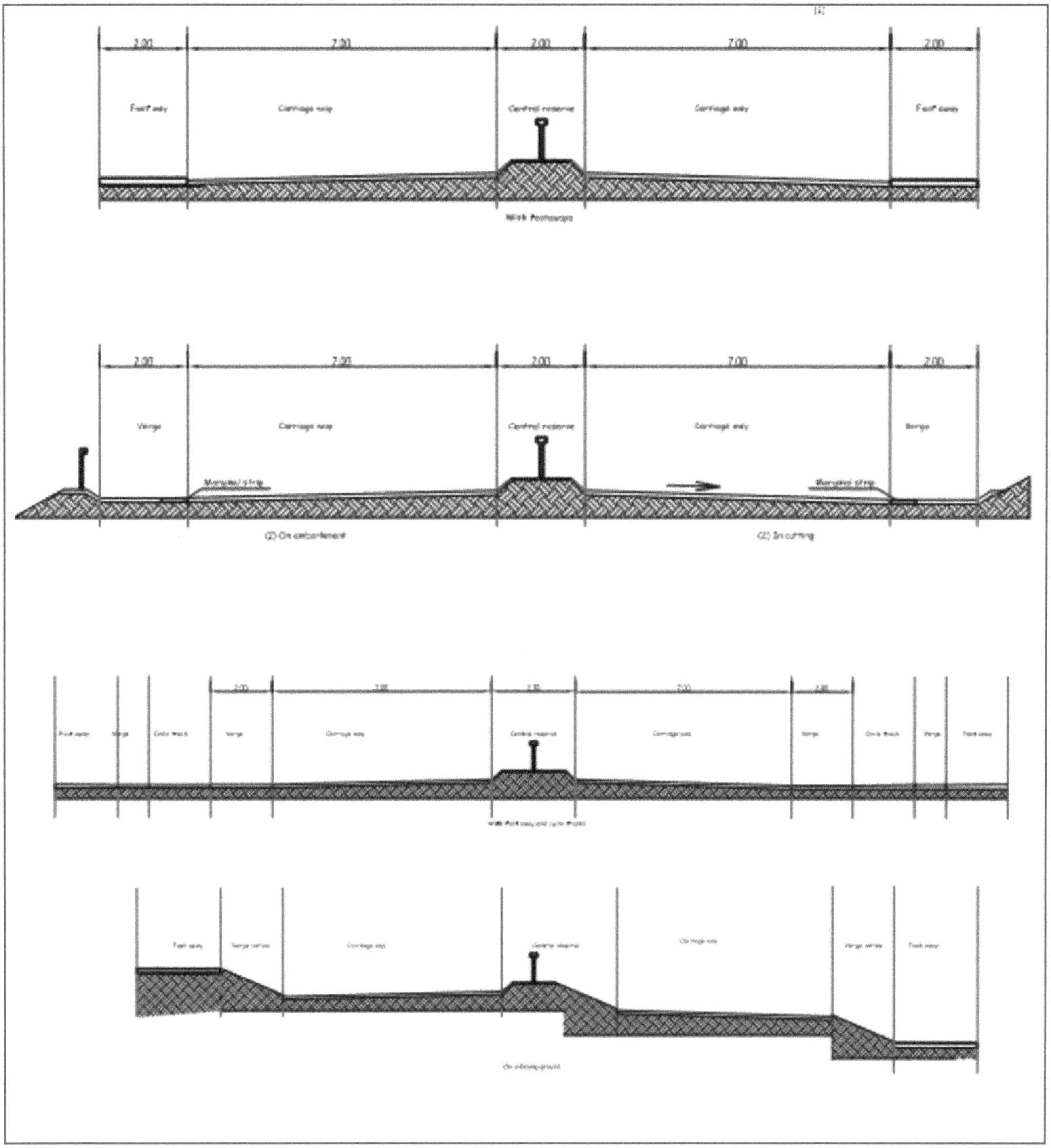

The City Street in Camaroon

Photo 29. Air Afrique Junction, Douala.

Photo 30. Uncontrolled parking, Bafoussam.

Photo 31. Street degradation.

Photo 32. Channelling of the Mfoundi River, Yaoundé, destroyed mosquito habitats.

Chapter 14

Photo 33. Damage caused by leaking water pipes under the carriageway. No timely repairs.

Photo 34. Lack of building codes and norms; no codes of conduct; safety risks.

Photo 35. Collapsed building, Santa Barbara, Yaoundé; no site control by engineers.

Photo 36. Questionable structural design; improvised scaffolding; no security measures.

75

The City Street in Camaroon

    To this end, the owner of the project would commission a study with appropriate terms of reference to define the technical parameters of the street. With this particular project (the rehabilitation of *Ancienne Route* in Bonabéri, Douala), there is the impression that the study (if any was carried out) failed to understand the classification and function of the said street as a transit route. It also failed to take into consideration vital factors such as population trends over the years, employment trends, an origin and destination survey, land use and, above all, the current traffic count on the street.

    A detailed study of the above issues would have determined the overall movement of people and goods within the influence zone of the project. In addition, historical information about traffic (from Douala municipal authorities) should have thrown light on the fact that the street, popularly known as *Ancienne Route* because it was constructed long ago, together with its sister tributary, often referred to as *Nouvelle Route*, are the transit sections of National Highway 5 (Douala–Bekoko–Tiko; Douala–Bekoko–Bafoussam). These two streets carry a combined ADT of 11,680[12] vehicles (based on the 2010-2011 traffic counts) – the densest traffic in the country. This should have been taken into consideration in the street design for rehabilitation. Continuity of routes is essential; otherwise, through traffic movements will be retarded. Transiting Douala from Yaoundé was becoming increasingly difficult.

    Experience in traffic management obliges the design engineers to ensure that street rehabilitation reflects the existing function of the street by providing a level of service superior to that which existed before the work began. Any reduction of the geometric characteristics of the street would only lead to additional traffic congestion. Unfortunately, the rehabilitation of *Ancienne Route* in Bonabéri-Douala made it smaller than it was before rehabilitation.

    A suggestion is being made to the authorities that these two streets should be transformed into one-way streets with dual carriageways right from the Bonabéri Bridge to Bekoko Junction. This would provide efficiency and reduce both the incidence of traffic congestion and traffic accidents from head-on collisions (see Diagram 9). Traffic statistics indicate that the ADT flowing into and out of Douala is on average 11,680 vehicles. Consequently, a more appropriate solution would be the construction of a 4x4 divided carriageway because over-taking has become virtually impossible on the existing two-lane streets (see Table 10).

---

[12] (Source: Ministry of Public Works – road traffic census 2010-2011)

Diagram 9: Proposed modification for traffic circulation in Bonabéri, Douala

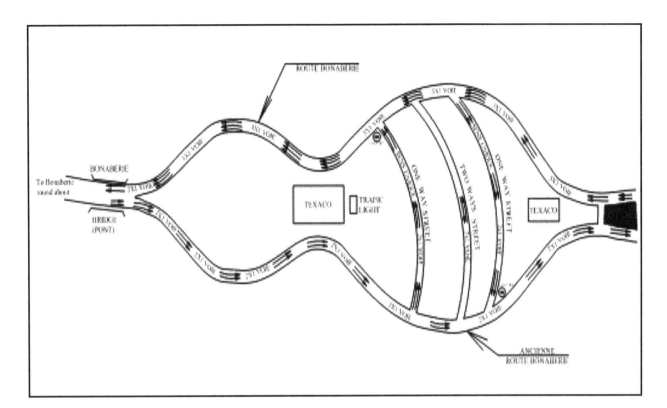

14.3 Pedestrian walking distances

Sometimes society requires that design engineers and planners should provide answers to abstract psychological questions such as, what is the optimal distance people will tolerate walking before they catch a taxi or bus? Most of the chaos on our streets can be traced to our inability to provide adequate answers to this question. Even when approximate answers have been provided by signposting bus and taxi stops, some people still ignore the signs, preferring to be picked up by a taxi wherever they are on the street.

Another question concerns how far motorists are willing to drive to find a free parking space instead of paying for an allocated parking spot. These are personal behavioural trends and individual preferences which further complicate street design. Sometimes it is almost impossible to establish the appropriate mix between what people do for their own convenience and what society requires of them for the smooth functioning of the community. Perhaps engineering schools and institutes of behavioural sciences should begin guided research to find answers which will enable future planners to understand the mindset of street users.

CHAPTER 15
Bypasses

If there were records to show the effects or impact of applying certain strategic options in city and street planning, there would by now be a planning policy that provides bypasses whenever a major highway becomes a transit route through a city or other urban centre. The main objective of providing these detours or bypasses around population centres is to limit traffic congestion to a minimum within the urban centres. Other objectives are to reduce travel time for transit traffic, reduce traffic accidents, and reduce the attendant pollution and environmental degradation associated with allowing through traffic to use urban centres.

The existing bypasses in Nkongsamba, Bangangté, Bafang, Bafia, Victoria, and Yaoundé provide evidence to support the objectives enumerated above. If an origin-and-destination study were to be conducted at Bekoko for vehicles coming into Douala from the Tiko direction, and those from the Loum-Mbanga direction, it would be discovered that there is a percentage of traffic that does not want to stop in Douala. It is this traffic pattern that very often justifies bypasses. The same scenario can be repeated for traffic travelling into Douala from the Edéa-Yaoundé Highway with similar results.

15.1 Creating the Bekoko Junction bypass

Creating a bypass to link Bekoko Junction on National Highway 3 to the Douala-Yaoundé Highway was to be a long-term engineering solution to traffic congestion. In 1985, a meeting in the Ministry of Planning and Regional Development was held to examine an offer made by JONIFEL (an international company) to pre-finance the construction of a new eight- or four-lane bridge further upstream on the Wouri River to link the Douala-Yaoundé and Douala-Bekoko highways. Work was to include feasibility studies, detailed design studies, and construction.

Research findings indicate that a second bridge on the Wouri River was retained in both the fourth and fifth five-year government development plans – the blueprints for government action concerning the socio-economic development of the nation. That bridge was never built. The current proposal coming 27 years later must be appraised from the historical background of a long-delayed project. The exact locations of the beginning and end points will be determined after conducting feasibility studies. The proposal trajectory envisages that the second bridge should be constructed upsteam on the Wouri River before it enters the marshy lands and breaks up (Diagram 10). The bypass would cross the old Douala-Edéa road before merging into the present Douala-Edéa-Yaoundé Highway (somewhere before the existing Dibamba Bridge on the Douala side).

The main bypass should be an eight- or six-lane separated (dual) carriageway fenced in to avoid intruders building along it on the expropriated land, thus transferring the traffic congestion to the bypass. The highway should have a central island separating the lanes, and should be lighted. Traffic counts of 11,680 vehicles for Bekoko-Douala and 5,036 for Douala-Edéa (Table 10) justify the creation of this bypass.

Diagram 10: Proposed Douala bypass

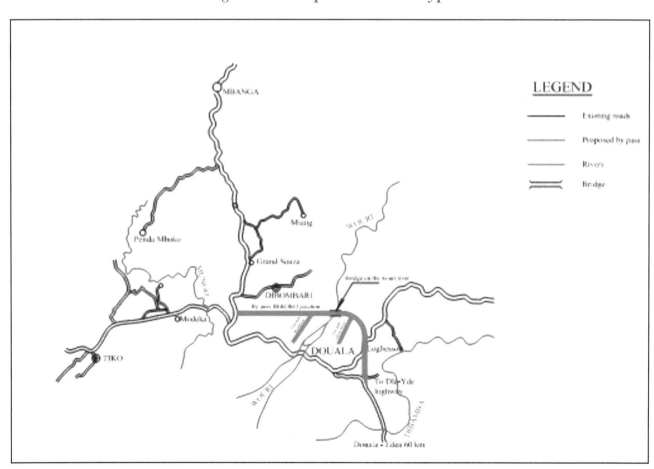

The second bridge should be designed to carry eight lanes even if the bypass retained is only six lanes. Provision should be made for the laying of utility service lines (electricity, water, telecommunication, etc.). At entry and exit points, the number of lanes should be doubled to facilitate toll collection and limit any congestion to a strict minimum.

Depending on aerial photography and the possibilities offered, secondary streets may be created to link the existing streets in the neighbourhood to the bypass to further decongest the city of Douala (Diagram 10). The plan also relays traffic from the city to the bypass on both sides of the proposed new bridge upstream on the Wouri River.

Creating the bypass will have various advantages:

(a) Transit traffic from both sides will no longer be held up in traffic jams within Douala municipality;

(b) Time savings, reduction in traffic jams, reduced traffic accidents, and attendant pollution associated with traffic congestion;

(c) Improved business opportunities arising from the availability of prime land at cheaper prices following the creation of the bypass;

(d) Increased property and land taxes will provide additional revenue for the city;

(e) Tolls from the bypass will attract investors to take on the project on a build, operate and transfer basis;

(f) Revenue can also be generated by selling rights-of-way to utility service providers who wish to install or lay lines or pipes;

(g) The construction of new social infrastructure (schools, hospitals, health centres, markets, etc.) will generate employment and revenue from taxes;

(h) New industrial areas will attract new businesses;

(i) Revenue from advertisements on billboards signposted on the bypass will increase the budget of the city council.

15.2 Additional bypasses

Economic growth tends to induce traffic growth as well. Matthew Edel and Jerome Rothenberg have observed that "the distinction between transportation and locational behaviour, between short-run and long-run effects, is the crux of the difference between urban transportation planning as an engineering exercise on the one hand, and as the design of a framework of interaction for a viable urban community, on the other" (*Readings in Urban Economics*, p. 491). In the short-run, we shall live with the traffic congestion, but in the medium-term our plans will be adjusted to handle new pressures.

The City Street in Camaroon

The city street must be concerned with prospective events that have impacts on the functioning of existing streets. The upgrading of the Ekok-Mamfe-Bamenda highway into an international road (Trans African Highway) means more traffic flowing into the city centre and more congestion. Bafoussam will need two bypasses to avoid saturation of the city centre. The first one will link the former Dschang Road junction to the highway coming in from Bandjoun (Diagram 11). The second bypass will link the Bamenda-Bafoussam Highway beyond the Bafoussam Airport to the Bafoussam-Foumban Highway (the exit route of the future Trans-African Highway that runs north through Banyo to Tibati-Meiganga and Garoua-Boulaï.

As already explained, the Trans-African Highway will enter Bamenda through Bali. An appropriate bypass should be designed to link this highway just before the City of Bamenda to the Bamenda-Bafoussam Highway at some point near or after Km 13. This would avoid the dangerous hillside that has often provoked landslides resulting in closures of the road (Diagram 11). It should be noted that even in Bali, a bypass will be needed because the present situation cannot be sustained for long.

Diagram 11: Proposed Bamenda and Bafoussam bypasses

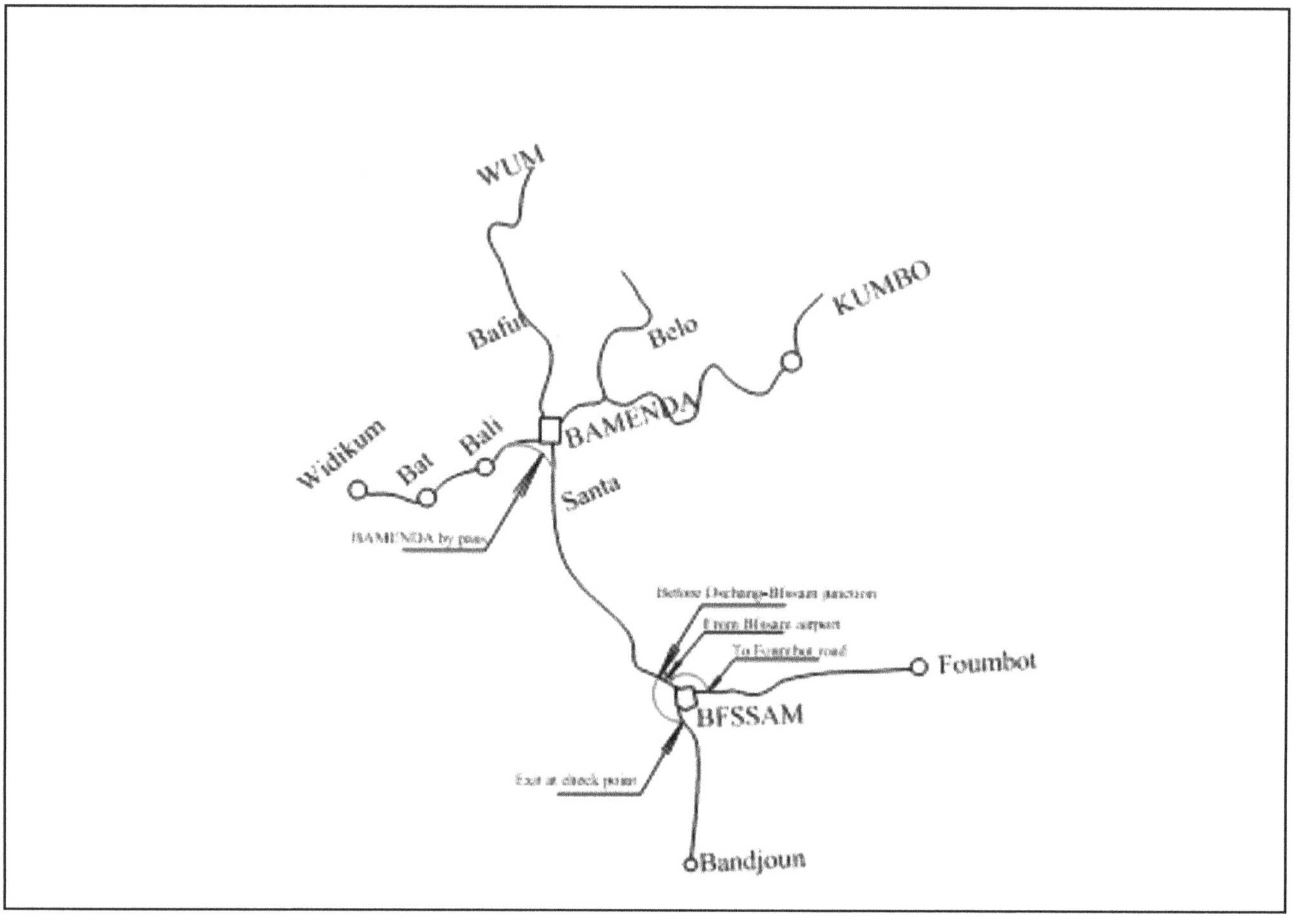

CHAPTER 16
Building construction on city streets

The street is like a river confluence where everything (business activities, housing construction, traffic circulation, pedestrian movements, hawkers, etc.) seems to gravitate towards one area. Because building construction has such an important role to play in city street planning, it must be discussed as a primary concern when dealing with streets. This construction begins with the provision of shelters as well as the formation of shops, factories, banks, business premises, schools, hospitals, ministries, courts, pharmacies, etc.

This section of the book will concentrate not only on building construction but primarily on the materials for construction and how they should be used to guarantee the safety of the occupants of the structures built and the stability and durability of the buildings erected. Professional societies regulate activities in the building construction sector for a variety of reasons. To begin with, specific construction techniques must be adopted in conformity with local materials and conditions.

In order to ensure uniformity in the application of construction techniques, certain specifications are issued by national or other widely-recognized bodies. The general rule is that these specifications are defined and referred to in contract terms as *standard specifications*. They are often represented by abbreviations as follows:

(a) BS – British Standard;

(b) BSCP or CP – British Standard Code of Practice or Code of Practice;

(c) AASHO – American Association of State Highway Officials;

(d) AASHTO – American Association of State Highway and Transport Officials.

16.1 Specifications for building materials

The application of standard specifications to construction work is governed in principle by the adoption of test procedures applicable to specific construction materials. The test procedures applicable to stone, sand, and filler are shown in Table 11. Those applicable to cement and lime-treated materials (based on foreign norms) appear in Table 12. Unfortunately, despite all the pressure mounted by the Cameroon Society of Engineers (CSE) through the Ministry of Commerce, Cameroon builders are still waiting for official local specifications to be used in construction. It is instructive to mention that ministerial orders or decrees alone cannot substitute for proper standard specifications. Every effort must be made to encourage builders to construct in accordance with the highest levels required by the specifications to ensure structural stability, safety and durability.

Table 11: Examples of test procedures applicable to stone, sand, and filler

Determination of	Test procedure
Particle size distribution	BS 812
Clay, silt and dust in aggregate	BS 812
Flakiness index	BS 812
Relative density-water absorption	BS 812 – part 2
Bulk density, voids and bulking	BS 812
Moisture content	BS 812 – standard method (oven-drying)
Aggregate crushing value	BS 812
Soluble chloride content	BS 812
Organic impurities in sands	AASHTO T21
Los Angeles abrasion	AASHTO T96 (ASTM C 131)(coarse aggregate)
	ASTM C 535 (large size coarse aggregate)
Sodium sulphate soundness	AASHTO T104 (ASTM 88) (5 cycles)
Sand equivalent	AASHTO T176 – mechanical shaker or manual shaker method

Ordinary and rapid-hardening Portland cement should be sampled and tested in accordance with all the requirements of BS 812. Building lime should be sampled and tested in accordance and compliance with all the requirements of BS 890 (Table 12).

Lime for treatment of road materials should be hydrated calcium lime or quicklime and, unless otherwise specified, should comply with the following requirements. The fineness residue of hydrated calcium lime on a 0.2-mm sieve should be a maximum of 1%; that for quicklime should be a maximum 10%. The fineness residue of hydrated calcium lime on a 0.0275-mm sieve should be a maximum of 10%; that for quicklime should be a maximum of 50%.

Sampling of cement or lime-treated materials should be carried out as specified or instructed by the engineers. Samples should be prepared for testing as indicated in Clause 1.5.3 of BS 1924. However, samples containing particles larger than 20 mm shall be prepared for compaction and CBR tests as indicated. Fraction coarser than 20 mm should be rejected but replaced by an equal weight of 5/20 mm material). Standard methods of testing cement or lime-treated materials should be performed in accordance with those given in Table 12.

Table 12: Examples of test procedures applicable to cement and lime-treated materials

Determination of	Test procedure
Moisture content	BS 1924 – Test 1 (A), oven drying method
Density-moisture content relationship (2.5 kg rammer)	AASHTO T99 – Method A (materials with not less than 90% passing 5-mm sieve)
	AASHTO T99 – Method D (materials with less than 90% passing 5-mm sieve)
Density-moisture content relationship (4.5 kg rammer)	AASHTO T180 – Method A (materials with less than 90% passing 5-mm sieve)
	AASHTO T180 – Method D (materials with less than 90% passing 5-mm sieve)
Density-moisture content relationship (vibrating hammer)	BS 1924 – Test 5
Unconfined compressive strength	BS 1924 – Test 10 (fine and medium grained materials)
	BS 1924 – Test 11 (medium and course grained materials)
Effect of immersion on UCS	
California bearing ratio	BS 1924 – Test 12
Cement content Lime content Field Dry Density	BS – Test 13 Dynamic compaction Method 1, soaking as AASHTO T193
	BS 1924 – Test 14
	BS 1924 – Test 15
	BS 1924 – Test 6 or Test 7
	AASHTO T238 – Method B, moisture content as BS 1924 Test 1 (A)

16.2 Guidelines for concrete construction

Concrete is the most visible and most extensively used material in building and civil engineering works. The quality of concrete is influenced by respect for the specifications of the materials, the procedure adopted for concreting, and the care given the concrete once it has been put in place. In simple technical language, concrete consists mainly of fine aggregates, course aggregate, water and cement as binders; all of these must be combined in optimal proportions to obtain the best product.

The design of reinforced-concrete is executed in accordance with the specifications of Building Code Requirements for Reinforced Concrete of the American Concrete Institute (ACI). The ACI reinforced concrete design handbook contains many useful tables that expedite design work (Tyler G. Hicks, *Standard Handbook of Engineering Calculations*, pp. 1-122). The spacing of steel reinforcement bars in the concrete is subject to the restrictions imposed by the ACI Code (see the reference books mentioned above). The quality of the concrete used for construction largely determines the ability of the building to withstand shocks from external forces.

A suitable concrete mix relies on the choice and proportion of the right ingredients used; a concrete mix should be economical to produce and should have the desired properties both when fresh and when hardened (that is, cured under appropriate conditions).

Fresh concrete should have both workability and consistency. Workability refers to the ease with which concrete can be mixed, handled and placed without segregation. Consistency refers to the fluidity or degree of stiffness of the fresh concrete and is a component of workability. In order to control the properties of workability and consistency, a fresh concrete mix should be tested using a truncated cone form. The cone is filled with concrete following a prescribed procedure. It is then withdrawn, and the fall or slump of the concrete is measured. This test is commonly conducted in the field as a means of controlling the quality of concrete.

Workability and consistency are influenced by the various components in the mixture and their properties. Quality concrete depends on the control of the following variables: water:cement ratio, maximum aggregate size, aggregate grading, aggregate:cement ratio, and the use of admixtures. As for the aggregates, they must be hard enough to resist abrasion and sound enough to withstand repeated cycles of rain and sun (if exposed). Aggregates should also be strong enough to produce concrete that has compression and flexure; they should be tough enough to resist impact and wear, and they must be inert to chemical reaction with the alkaline cement.

The addition of excess water to the concrete mix makes it more fluid, but it also makes the concrete weaker, less durable and less water-tight. The size and gradation of the aggregation of the aggregate is very important; therefore, no cracked stones should be used because they have no defined sizes. A majority of the concrete used on private sites cannot pass a slump test due to the overdose of water. The water:cement ratio has been the most abused and disrespected variable on local construction sites. This situation will not change until the national Order of Civil Engineers does something to intervene.

Not long ago, concrete had only to satisfy compressive strength as a main parameter. Today, concrete has also to meet other durability parameters such as least levels of water penetration, rapid chloride permeability, water absorption, etc. The international standards BSEN 12390-8, BS 1881-208, ASTM C1202, ASTMC 109, BS 6319-3, etc. govern the performance of fresh and hardened concrete. Engineers on construction sites should take note and pay particular attention to the requirements for modern concrete mixtures.

16.3 Building foundations

Engineers are designers, and design, engineering, and business form a three-party activity. The emphasis here is on design engineering because foundations of buildings need to be adequately designed. It has become common practice locally to have draftsmen and non-engineering professionals take it upon themselves to draw plans and construct houses without having the proper certification as design engineers. This is not only prejudicial to engineers; it is against the law because such activities put the safety of the public at risk.

It has been observed that the single most common cause of failure in building masonry is the separation of walls at corners because they were not adequately tied together during the design of the building. In designing the foundation, care must be taken to ensure that horizontal reinforcement ties up the walls, floors and the roof plane together with a ring beam. Soil investigations should be carried out to ensure that the ground is stable enough to transmit loads evenly to the layer beneath. All the elements of the foundation should be well-connected to spread the load from the building into firm ground and to prevent cracks resulting from uneven settlement.

Economic hardships and the scarcity of land within urban areas are forcing people to revisit their investment options. There is a trend slowly taking root in the housing sector whereby people are carrying out modifications or extensions to their existing houses while ignoring structural symmetry. Others are quietly transforming single floor buildings into storey buildings without any corresponding redesign of the foundation: they simply add one floor on top of another. The danger is that spectacular collapses may occur due to surcharge loads or inadequate foundations. Something must be done to avert the loss of lives that may result from such collapses (see Photos 34, 35, 36).

The other danger that is being overlooked is the uncontrolled construction of houses in earthquake-prone areas. In Buea, Limbe, and Ekona, houses are mushrooming because of the demand brought about by the creation of university institutions and their affiliates. It should be emphasized that those investing in the provision of houses in those areas must consult before building. There are many reasons for this. An earthquake sets off internal forces in a building which cause its elements to distort, twist, stretch and bend in ways which are quite different in each earthquake and therefore are difficult to predict. In eastern Turkey and Egypt (1992), many multi-storey buildings collapsed because they were not designed and constructed to resist the intensity of the shaking which they experienced

(*Technical Principles of Building for Safety*, p. 57).

Because it is not currently possible to predict an earthquake with any accuracy, the authorities must make it mandatory that all buildings more than one floor should be subjected to the application of robust seismic design safety regulations. These factors are found in seismic design codes and are meant to be used for the structural design, stability and safety of buildings. Neglect in applying seismic design codes when building must be seen as endangering human lives. Reinforced concrete structures that are wrongly designed or poorly constructed are quite vulnerable, and in the event of a collapse they are very lethal because of their heavy weight (*Technical Principles of Building for Safety*). The best building principle to follow is to use robust design, good foundations and adequately supervised construction.

16.4 Building permits

Building permits are routinely signed and issued to those who have applied and paid for them. A building permit is granted by the city council after the applicant has met certain requirements. The entire procedure is modelled after the colonial administration to reflect that practised in Europe. What it does not take account of is that city staff in Europe include town planners, engineers of various fields of specialization, architects, lawyers, environmentalists, etc. In Cameroon city authorities include few professional and technical staff. There is a deficit of engineers and other technical personnel.

Concerned citizens have questions about building permits and sites. For example, what are the responsibilities of the city council towards their clients as concerns adherence to construction norms (the foreign ones currently in use)? Who determines whether approved plans meet technical specifications for construction or sanitary regulations governing the disposal of waste water? Who enforces safety and health regulations on site? During construction, are public goods (streets) and public utility services protected?

Owners of large buildings that occupy space in the city should provide parking space in conformity with the parking norms linked to new housing construction. There should be provision of one parking space for 40 to 60 m^2 of built up office space. For buildings in the city or town centre, there should be one parking space for 80 to 100 m^2 of built up area. It is imperative that the parking space allocated should be shown on the plans approved. A question arises as to why the planned and constructed three-storey parking ramp attached to the 18-storey ministerial building on 20th May Boulevard (Yaoundé) was abandoned. If the Government cannot manage the parking, it should be privatized to recover the taxpayer money used to finance the construction.

The absence of regulations in the housing construction sector means that anyone can decide to build anywhere at any time. Many people seek no advice about financing or engineering, resorting to draftsmen instead. Soon their bank accounts dry up and the buildings go unfinished or are abandoned. These structures sour city landscapes and pose safety and health hazards. The question remains: how long should abandoned buildings be allowed to stand unfinished? If at some stage the owner decides to restart the work, what conditions must they fulfil to guarantee the stability and safety of the building (before a new permit is issued for work to resume)?

16.5 Construction cost estimates

Field experience has led to acceptance that some of the parameters used in project evaluation are subject to two main sources of error. Construction costs, for example, contain errors associated with the measurement of quantities and price estimates, even with such things as assumed foundation conditions (at the design stage). The availability of raw materials can also be factored into the equation. Dealing with the measurement of quantities for huge projects can sometimes be a daunting task. Engineers should therefore combine their expertise with that of other specialists such as quantity surveyors (whose specialized knowledge can help limit errors in estimation).

Unfortunately, there are very few quantity surveyors in the country. As for foundations, their design should incorporate any additional soil investigations or open excavations that reveal the nature, type and content of the foundation material at the construction site. An engineer can give a fairly definite range of building costs followed by a wider range to cover unexpected site conditions, etc. This is a more natural form for presenting engineering cost data since the standard contingency allowance is nothing more than an explicit allowance for certain types of uncertainty.

Building cost estimates should be updated in the course of project execution if and when cost parameters change. This is because they are, after all, only estimates, and their real purpose is to enable the owner of the project to carry out a complete plan of construction. Construction management requires detailed plans of execution (supply of materials, payments to the workforce, supervision of loan agreements, etc.). Cost estimates, above all, provide the owner with information on the finances needed for construction.

CHAPTER 17
Enforcement of norms and regulations

Shelter has become a principal preoccupation of human beings. It would appear that the sky is the limit as to what can be designed and constructed. However, the major concern that architects and engineers have is how to fashion buildings to match local conditions and environments. The alternative is simply to import designs that are popular elsewhere.

An example of an inappropriate or "out-of-place" design is the Ministry of National Education in Yaoundé. The roof was designed as a collector instead of a discharger of rainwater. Huge concrete slabs covered with waterproof material was used instead of corrugated roofing sheets. In the tropics where there is heavy rainfall, run-off water should be discharged as quickly as possible. The building was also fitted with imported glass doors whereas wooden doors could have been manufactured locally. Further, the building was designed to be solely reliant on central air conditioning with no concern for the maintenance costs of such equipment. Lifts were installed to be powered by electricity which was very often irregular. Even when standby electrical generators were supplied, they had to rely on the availability of technicians, generator fuel, and spare parts.

The building is still suffering from the absence of cross-ventilation design. Other problems remain: central air conditioning units do not work, water pumps have failed, and lifts seldom work. Most of the lifts were forced to operate against their automated system and soon became useless. After only a few years, the central air conditioning system failed completely and toilets slowly degenerated into disuse.

A three-storey parking facility was constructed and attached to Ministerial Building No. 2 in downtown Yaoundé. At the time of commissioning, the 300-vehicle parking lot was functional and adequately signposted inside and out with adequate directional signs and complete lighting inside the structure. Today, this facility lies abandoned while vehicles are parked along the streets and sidewalks. Who then is responsible for the proper use and maintenance of public buildings? And what possible explanation can be given to justify such waste of public resources?

17.1 New regulations

The above review of building construction acknowledged the fact that there has been borrowing of norms and standard specifications from Europe or America. Although the above norms are known and quoted in the terms of reference of construction contracts, their application by contractors or private individuals has been largely symbolic. This lack of control and enforcement has resulted in an increase in the number of collapsed buildings.

Fortunately, the Government adopted Law No. 2000/09 of 13 July 2000 relating to the organization and practice of the profession of civil engineer. It states unequivocally in its general provisions (Part I, Section 3.1) as follows:

> The practice of the profession of engineer shall concern any creative activity or work requiring the training and experience of an engineer in civil engineering specialities, with a view to ensuring that the said creative activity or work conform to specifications and plans, and complies with the rules of the trade. This applies to construction works, installation works, buildings and, in general, to projects for the construction of public and private infrastructure.

In Section 37(1), the law further states:
> Pursuant to the provisions of Sections 24 and 25, the council of the order shall, as the institutional partner of the public authorities, contribute in the formulation of strategies, for decision-making and for the implementation of policies in sectors needing civil engineering expertise.

Responsibility for enforcement of construction norms, the code of practice, standards and regulations in the housing and other civil engineering construction activities, lies squarely on the National Order of Civil Engineers (NOCE). The immediate task should be to draw up guidelines that city councils will comply with when issuing building permits for housing and other related construction works. Such guidelines must include the introduction of site inspections by engineers for large projects and provision for a project control engineer by the project owner for ongoing supervision. The guidelines should make it mandatory for the names of site engineers to appear on the project construction billboards at all construction sites. Compliance should be monitored by council inspectors, and those who violate the law should be taken to court to justify endangerment of public safety in their construction sites.

Because the foreign norms and standards previously adopted are based on the properties of their local materials, the National Order of Civil Engineers should intensify the drive towards the establishment of norms for local materials (sand, cement, water, wood, stone aggregate, soils, etc.). These local materials have their own properties not necessarily inherent in the foreign norms.

Unfortunately, some engineers have an ambivalent attitude which borders on absolute indifference and carelessness when it comes to technical responsibility on issues of public interest. The collapse of the bridge over the Mungo in 2004 was met with undignified silence from the National Order of Civil Engineers. There was no officially-organized visit to the site to record first-hand what had happened. Again, when private buildings collapsed in Douala, Yaoundé and Bafoussam, the NOCE should have set up a technical commission to inquire and report why the buildings collapsed (once again, there was no public reaction).

It is immediately emphasized that such indifference from engineers is incompatible with the mission of engineer outlined in this book. It is unheard of for engineers to remain silent amidst the chaos in our streets. As evidence of collapsed buildings continues to grow, it only reinforces the role of NOCE to regulate and control. In 2013, *The Post* reported the collapse of a six-storey building on Mermoz Street, Akwa, Douala killing two people. It was revealed that city authorities had been contacted by neighbours, to no avail, to inspect the work and take appropriate safety measures. The magistrates' court at Bonanjo had issued a court order suspending the work, but this was ignored by the owner (*The Post*, 22 July 2013, p. 9).

Prior to this, a one-story building collapsed at *Afrique du Sud,* Bonamoussadi, Douala, while in May of the same year, a building collapsed at Mobil-Guinness, Ndokoti, Douala (note the popular as opposed to official street names). NOCE reacted by requesting reports about these collapsed buildings from the local authorities.

Engineers have the right to examine, analyse and control the implementation of construction activities. That right has been enshrined in law. There is no reason why engineers cannot halt the spread of collapsing buildings and force out the charlatans who are perpetuating the chaos in the building industry. The responsibilities of NOCE have just begun. Is there a building policy for public office buildings? Why are office buildings not limited in height to four or five storeys. Why is effective cross ventilation not used instead of central air-conditioning? What allocations should be made for the maintenance of government buildings?

17.2 Abuse of city streets

By its very nature and definition, a city street is a public good subject to the principle that consumption by an individual should not diminish consumption by others. Consequently, any action or activity carried out on the street that impedes the attainment of the said objective constitutes abuse.

Research findings reveal that in Cameroon, city authorities have refused to apply the now universal principle of engineering construction standards and norms applicable to the location of underground utility lines (these should be buried under the sidewalks at specific depths (see Diagram 5). Their refusal may stem from ignorance or an inherent tradition of rejecting change; however, it is causing tax-payers billions of francs in street destruction and subsequent repairs. The recent destruction of a street in Tsinga and other locations in Yaoundé could have been avoided (Photo F in the Annex). The picture bears testimony to the fact that the street was built to last for a long time (note the thickness of the asphalt concrete pavement and the base course).

Sidewalks on city streets were elevated and set aside to separate pedestrians from vehicle traffic. Unfortunately, in some areas of the towns and cities, sidewalks have been converted to parking places. This has pushed pedestrians from the sidewalks onto the carriageways. This problem should be urgently addressed.

The catalogue of abuses committed against the street are long and compelling. Only the most glaring examples are discussed here.

(a) In Cameroon, driving is on the right, but on the left inside the roundabouts. However, exiting the roundabout from the inner lanes is a violation of the driving code. When the sidewalks designed for pedestrian traffic are transformed into call-box centres, kiosks for selling lottery tickets, telephones, and merchandise (clothes, firewood, fruits, vegetables), this is abuse;

(b) When sidewalks are transformed into kitchens and eating places, or when they are used as garages for car and truck repairs, this is abuse;

(c) Another example is the way in which ditches have been converted into other uses. These were provided as drainage structures for surface water run-off. They have been converted into dust bins, refuse dumps, and waste water disposal channels from individual private buildings;

(d) Taxi drivers are known to deliberately ignore "No Parking" signs and block traffic to pick up passengers, or else form two parallel queues instead of one at taxi ranks, thus causing congestion; this is abuse of the street. They also habitually refuse to queue at taxi ranks, preferring to line up parallel to one another to pick up passengers, thereby creating traffic jams;

(e) When vehicles are parked right up to the traffic light (such as in Melen, Yaoundé), motorists are unable to move freely.

(f) When whole streets are converted into marketplaces by street vendors (as in Mokolo, Melen, Mfoundi, Mvog-Mbi, etc., in Yaoundé), this is abusive use of the street;

(g) When motorcycles drive on centre islands in order to beat traffic congestion, it is an offence (this offence is often overlooked);

(h) Perhaps of all the abuses, the one that involves almost all citizens is littering. There are two types of littering: voluntary and involuntary. Voluntary littering is when people deliberately drop unwanted waste on the street. An example of involuntary littering is when customers become agents through the use of plastic bags to wrap items sold. Many countries have banned the use of plastic bags by retailers. Other countries are still sitting on the fence, not knowing what to do about this public nuisance. Italy has advocated a total ban. A vast programme of civic education at all levels of society is needed to reverse the indifference to dirtying the street.

Chapter 17

In Cameroon, littering is so widespread that one wonders what people do in other countries to keep their streets clean. The most common litter includes sugar-cane chaff; groundnut shells; maize cobs; banana, mango and orange peels; scrap paper; and a host of other trash. These are used and dumped by pedestrians and motorists almost anywhere in the streets;

(i) The advent of plastic bags has compounded the problem of littering. Villages as well as beautiful rural and urban landscapes have been defaced by discarded plastic bags. These bags fly about, float around, and become trapped in trees, shrubs, ditches and waterways. They block drainage structures and have even caused flooding;

(j) Furthermore, there is the million-dollar question: What is the speed limit on city streets? Nowhere are there any signposts indicating the speed at which motorists should drive in towns and cities. This is dangerous and unsafe for urban populations. For example, in January 2007 at the street junction entering City Bilingual College in Etoug-Ebé, Yaoundé, a truck descending with speed from the direction of the Etoug-Ebé Handicapped Centre in Yaoundé suddenly lost control, climbed the sidewalk and knocked down three children who were returning from school. One died on the spot and the other two were rushed to the hospital. Today, there is still no preventive measure to control speed on this dangerous descent: no speed brake, and no speed limit. What happened to civic responsibility? Why are there no efforts to prevent these accidents?

(k) When poles carrying telephone and electricity cables fall onto the street and are not immediately replaced by Camtel or AES-Sonel, this constitutes abuse of the street. Again, it is negligence on the part of city authorities who should summon these companies and compel them to act urgently to restore the normal functioning of the street;

(l) Furthermore, there is the million-dollar question: What is the speed limit on city streets? Nowhere are there any signposts indicating the speed at which motorists should drive in towns and cities. This is dangerous and unsafe for urban populations. For example, in January 2007 at the street junction entering City Bilingual College in Etoug-Ebé, Yaoundé, a truck descending with speed from the direction of the Etoug-Ebé Handicapped Centre in Yaoundé suddenly lost control, climbed the sidewalk and knocked down three children who were returning from school. One died on the spot and the other two were rushed to the hospital. Today, there is still no preventive measure to control speed on this dangerous descent: no speed brake, and no speed limit. What happened to civic responsibility? Why are there no efforts to prevent these accidents?

(m) When the water pipes buried under the carriageway burst, the water supply company (Camwater) digs up the street, repairs the pipes and leaves open trenches for months. These uncovered trenches accelerate the destruction of the pavement and cause traffic jams. This is a serious case of neglect and abuse of the street. It also represents a serious lack of responsibility on the part of city council authorities whose duty it is to ensure proper upkeep of the streets (Photo E, in the Annex);

(n) When poles carrying telephone and electricity cables fall onto the street and are not immediately replaced by Camtel or AES-Sonel, this constitutes abuse of the street. Again, it is negligence on the part of city authorities who should summon these companies and compel them to act urgently to restore the normal functioning of the street;

(o) When traffic and street lights are destroyed by reckless drivers who are not made to pay for the repairs, this is abuse of the street.

(p) It is improper for motorcyclists to create spontaneous motorcycle parks for themselves everywhere on city streets; for car dealers to invade and occupy city streets and sidewalks with vehicles for sale (such as in Akwa, Douala); or for people to occupy the city street with tents for funeral celebrations; these are all cases of abusive use of the street;

(q) When the authorities responsible for the management of city affairs refuse to enact meaningful legislation or pass laws to prevent these abuses, they too are guilty of abusing the street;

(r) When those who are responsible for ensuring strict compliance with traffic regulations and control devices fail to do their work, thus bringing about traffic congestion and even accidents, this is serious abuse of the street.

The list of abuses goes on ad infinitum because there is no legislation that has been enacted to make city street abuses punishable.

CHAPTER 18
Maintenance of city streets

There is a widely held view by the non-technical population that roads and streets once built should last forever. Unfortunately, nothing built by man is perfect. Everything that is conceived, designed and built has a defined lifespan contingent upon the system of maintenance adopted.

18.1 Reasons for good maintenance

It is imperative to state that all maintenance contracts must be awarded in time for their execution to be carried out completely within the dry season. Major rehabilitation work that requires the removal of all layers down to the foundation must never be done in the rainy season. In mid-2013, work was being done on the street running from the Etoug-Ebé junction towards Mendong in Yaoundé, and work was being done in Bonabéri before the entrance to Wouri Bridge in Douala. The subsequent failure of these projects was all but guaranteed because wet materials do not subject themselves to adequate compaction and therefore cannot acquire their optimum density. Also, water trapped inside the foundation, sub-base, base course and even in the pavements, eventually dries up leaving the structure porous and less durable. This explains why degradation soon happens as was the case at the entrance to the Bonabéri Motor Park in Douala. The contractors know about these shortfalls but they accept the work for the money, knowing that there are no penalties associated with subsequent failures of the infrastructure.

Street and road maintenance work must be carried out in conformity with the norms and standards of engineering practice universally accepted. Contracts for maintenance work must never be awarded to be executed in the rainy season. Those who award such contracts must be held accountable for the future disintegration of the streets and roads. Norms and standards concern the following:

(a) Where the structure is located as well as the forces of nature: rain, sun, weather, natural conditions of the environment;

(b) The limitations imposed on the design by the use of certain materials, choice of the materials, their handling, and placement of the materials;

(c) The construction method: equipment selection and its use during construction;

(d) The level of perfection in the execution by vigorous adherence to national or international codes of practice, or design standards, applicable to such engineering structures;

(e) The use of non-conforming materials: soils, aggregates, bitumen, cement, etc.;

(f) Methods used in the mixing, handling, placing, compacting and finishing of work.

Take the case of a contract awarded for the construction of a paved street. It is well known by engineers that the aggregate in a bituminous mixture supplies most of the stability. The mixture must also support the loads imposed and transmitted to the sub-base at reduced intensity. Unfortunately, some builders sometimes forget that the aggregates used in bituminous mixes tend to be broken or degraded by the loads imposed upon them, both during construction and later by the action of traffic. Another major cause of degradations is poor compaction during construction. This is especially true in the construction of surface treatment and seal construction.

It has been noted earlier that street traffic counts are virtually non-existent in cities in Cameroon. It is difficult to know, then, what determines street design and planning. Trends in the national vehicle fleet have also demonstrated that growth in the fleet has been very spasmodic, leading to uncontrolled traffic growth on the streets and heavier traffic loads. Therefore, two important variables must be considered: street traffic and the choice of construction equipment used by contractors. Lack of knowledge about these important variables requires that street maintenance be contracted at very early stages; this is preventive maintenance.

18.2 The value of city street maintenance

The returns on an economic investment depend to a large extent on the lifespan of the structure. Maintenance is the preservation and keeping of street infrastructure and facilities as close as possible to their original condition as constructed. This includes any subsequent improvement works and any additional work performed to keep traffic flowing safely.

Maintenance operations include patching of potholes and surfaces, removing surface corrugation, filling ruts, cleaning ditches and culverts, removing trash and litter, mowing or cutting grass and weeds, cleaning weep holes, erecting signs and traffic controls, painting traffic strips, re-enforcing traffic control signs, and removing broken down vehicles from the street.

Chapter 18

The end-product of adequate street maintenance is smooth, safe and efficient carriageways, clear waterways, and attractive streets and roads. Streets in different locations might require different maintenance operations. For example, those in Yaoundé, Douala, Maroua, Kumba, Bamenda, Bafoussam, Ngoundéré, Limbe or Kribi might reveal varying problems. In low-lying coastal regions with high water-tables, the maintenance of drainage structures is critical in order to avoid streets being washed away by floods. These structures must be kept clean and open at all times.

18.3 Organizing street maintenance

Effective street maintenance has to be organized and streamlined. The city council must determine whether to hire contractors or create maintenance crews based on the nature of the work to be done. Crews must also be equipped accordingly.

The choice of personnel and the level of their training is crucial for the success of maintenance operations. The group is normally placed under an experienced supervisor or foreman backed up by other workers. The number of workers should be proportional to the allocated task. Duties, responsibilities and line of authority should be clearly defined to avoid any conflict.

The type of equipment needed depends on the maintenance operations to be carried out. Equipment that is needed for only a short time could be hired to save costs, while other equipment might need to be purchased. It is always advisable to have detailed specifications drawn up for each major piece of equipment before calls to tender for their purchase are made.

The cost of buying equipment must be distinguished from maintenance costs. Maintenance expenditures include different aspects of work at different locations on the street. Appropriate records of accounts should be kept to reflect wages and the price of materials and equipment. For purposes of replacing equipment, there should be a depreciation account for each item. The funds needed for maintenance should be part of the city council budget.

Various activities are included in street maintenance as mentioned above. These are listed below without further details about the operations:

(a) Surface maintenance of laterite and gravel surfaces;

(b) Ditch cleaning on laterite and gravel surfaces;

(c) Routine patching on surface dressing;

(d) Routine patching on bituminous surfaces;

(e) Resurfacing of surface dressing;

(f) Resurfacing of bituminous surfaces;

(g) Cleaning of ditches (concrete covered or open);

(h) Cleaning of weep holes;

(i) Cutting of weeds on various widths of the street;

(j) Signing (of new and old signposts);

(k) Marking the centre line of streets;

(l) Marking no-passing lines;

(m) Reconstituting the structure of the street where it has been damaged during the repair of water pipes, electric cables, or communication cables, or from failures resulting from fatigue.

City street maintenance is an integral part of the planned development of a city. City budgets should include maintenance of existing streets, preventive work for those under construction, and rehabilitation of those streets whose life expectancy has ended. It is recommended, that city authorities should ensure that maintenance work is executed during the dry season only.

CHAPTER 19
The way forward

An idiomatic expression from Nwehland says that when a house caught fire, a rat ran out exclaiming: "What do I own in this house to allow myself to get burnt in it?" Unfortunately, citizens cannot run away from the problems created in cities, nor can they watch with indifference as the streets suffer and are subjected to all types of preventable abuses.

Movements of people and goods are at the heart of city life, and the city street is the nerve-centre of the socio-economic life of the community. Without streets there would be no cities. Burying our heads in the sand like the ostrich and pretending that street problems will go away on their own has not worked and will never work. It is not feasible to harbour the idea, like the rat, that when things get really bad, escape from the fire would be a solution. Unlike the rat, people own property in the cities and towns.

The only rational option is to confront these problems head on and find adequate and lasting solutions. Demographic statistics have indicated that in Cameroon the growth rate of the urban population was about 6% a year, and the rate of urbanization went from 28% in 1976 to 30% in 1987 (cf. Declaration of Government Urban Strategy, January 2000, approved by the Prime Minister and Head of Government, Letter B7/CAB/PM of 17 November 1999). These figures indicate the urgent need for serious urban and street planning. This book was written to document the abuses imposed on streets. Some of the abuses have actually distorted the functioning of streets.

The book has diagnosed the disease. Can the Government and the city authorities find the cure? The following approaches are suggested for the way forward:

(a) Carry out a detailed origin-and-destination study as a prerequisite for a better understanding of street traffic; subsequently classify all streets by function;

(b) Use both constructive and restrictive traffic engineering approaches to solve the problems of congestion and traffic accidents;

(c) Draw up a master plan of development for major cities and develop city or town maps showing existing and planned streets;

(d) Create traffic laws and ordinances (as suggested in this book) and have them passed into law by a competent legislative body;

(e) Create departments of traffic management within city and town councils and put them under the supervision of qualified traffic engineers (not civil administrators);

(f) Study, design, and construct bypasses for major cities;

(g) Develop and run traffic and driver education programmes on national and private television stations for the entire population;

(h) Draw up national construction norms and codes of practice which are based on local climate, materials and building techniques;

(i) Rigorously apply and adhere to seismic design construction in South-West Cameroon in order to avert disasters from future earthquakes;

(j) Protect the environment by halting the environmental degradation now caused by the illegal discharge onto streets of polluted water from buildings;

(k) Prepare and pass legislation regulating the disposal and treatment of toxic waste from filling stations and garages;

(l) Estimate future demand for public transport at the forecasting stage of all comprehensive land-use and transportation studies in all town and city development plans;

(m) Remove all utility service lines or pipes from under carriageways and transfer them to the sidewalks during any rehabilitation work;

(n) Review, evaluate and restructure the content of training programmes and teaching methods used in driving schools;

(o) Construct peripheral streets in Yaoundé and Douala (cf. the proposed bypasses for Douala and other cities);

(p) Renew and re-organize city and town markets with the possible option of relocating them away from major streets;

(q) Create five-year traffic plans for Yaoundé and Douala;

(r) Install traffic lights at major junctions and roundabouts;

(s) Formulate new policy guidelines limiting the age of imported vehicles (to a suggested 1 to 5 years maximum);

(t) Create an institute for training professional instructors for driving schools, and provide compulsory retraining of all existing drivers.

CHAPTER 20
Conclusion

Planning of city streets might be considered by some as an abstract subject. However, the physical presence of the street and its functioning are real. Others with more sinister views would consider such research work as an unnecessary and unproductive use of time. However, the study and research for this book were conceived and written to enlighten, educate and cause action to be taken by those in authority to improve street use and halt the abuses that threaten this public good.

The creation of an institution for the training of driving school instructors should be a priority. The requirements for operating driving schools should be reviewed and the period of training must be long enough to provide effective and professional education. A one-year course should include Civics and other related social sciences for professionals. Driver education for all should be presented in very simple language so that students of urban studies and the general public can understand and appreciate the city street as the nerve centre of urban activities.

In the wake of the last Japanese earthquake and tsunami, there was widespread speculation about the possibilities of a total evacuation of the city of Tokyo. How feasible would the exercise to evacuate millions of people from the city have been within so short a period of time? The answer will remain the subject of speculation for a long time. Records are still fresh about the forced evacuation of the citizens of New Orleans, Louisiana, following Hurricane Katrina and devastated floods. What lessons were learned and how should engineers use past disasters when designing future streets?

An educated guess is that as urban populations increase, engineers should begin the strenuous search for the model design of future streets. Perhaps it will be necessary to design and provide standby streets to be used in case of huge and rapid evacuations. Such streets, like unrented houses, might prove to be invaluable economic assets.

It is on record that the recommendations of the final report on the study of the Transportation Plan for Cameroon (Volume 3(a), pp. 49 to 51) have not been fully implemented. These need to be revisited, updated and implemented for the good of all.

Also included above in (n), (o) and (p) with slight modifications are the projects supposed to have been realized under the Sixth Five-year Socio-cultural and Economic Development Plan 1986–1991 (pp. 183–190). Their execution 20 years late would still alleviate traffic congestion and reduce accidents and environmental degradation.

Economics is about self-interest; this explains why the foreign policies of nations are influenced only by their perceived self-interest. The Bible would seem to agree because it states that "all achievement is founded on competitive envy" (cf. Eccles. 4:4). Some examples are in order.

After the Soviet Union successfully launched a manned spacecraft in 1963, the American president, John F Kennedy, vowed that America would land a man on the moon before the end of the decade. The task of execution was assigned to the National Aeronautics and Space Administration (NASA). America did land a man on the moon. Engineers conceived, designed and built the spacecraft which carried the men, but the government financed the project.

In the late 1960s or thereabouts, the Indian prime minister invited a core of Indian engineers from all over the world and challenged them to restructure the Indian Telecommunications sector. These engineers drew up a blueprint which their government financed. Today, India has become the number one nation for the outsourcing of information technology (revenue from it comes in the billions of dollars), while millions of people derive employment from the sector.

The president of South Korea undertook a similar exercise when he called upon engineers to design a model car. The government encouraged them by financing the project before ceding it to the private sector. Today, the Korean motor industry and their electronic products have become household names worldwide. They export millions of goods and create employment for millions of people. Engineers are the principal movers for industrialization and development as demonstrated above.

I would like to conclude this book with a quotation from Professor Bernard Nsokika Fonlon of blessed memory. He was an exceptional visionary, who wrote: "as the intelligentsia, whose role is of first importance in this enterprise, possess [but] skill without authority, they will not be able to give their best if the politicians persist in their failure to acknowledge the elementary principle that political authority, by itself alone, cannot be equated with technical know-how" (Fonlon in *Nation Building*, 1964). We should emulate the examples of the developing nations whose technical advancement was pivoted on the intellectual contribution of their engineers.

The City Street in Cameroon is an intellectual contribution designed to transform cities into the paradises they should be. After all, it is said that "each generation yearns to prove itself – and, in proving itself, to accomplish great things for humanity" (cf. Sherion B Nuland, 1994, p. 87). This book will remain a benchmark for the city street for generations yet to come.

Bibliography

Al Jazeera TV, The secret of the seven sisters, documentary aired 25 April 2013.

Bureau of Labor Statistics, *Engineers*, Washington, D.C.: U.S. Department of Labor, 2006.

Bureau of Labor Statistics, *Occupational Outlook Handbook 2006-07*, Washington, D.C.: U.S. Department of Labor, 2008.

CalEPA, Health effect of diesel exhausts particulate matter, Sacramento: California Environmental Protection Agency, 2007.

Coburn, Andrew, Richard Hughes, Antonios Pomonis, Robin Spence, et al., *Technical Principles of Building for Safety*, London: Intermediate Technology Publications, Ltd., 1995.

Department of the Environment, Scottish Development Department, *Roads in Urban Areas*, London: HMSO, 1983.

Downing, A.J. and I.A. Sayer, A preliminary study of children's road crossing knowledge in three developing countries, Crowthone Berkshire: Transport and Road Research Laboratory, 1982.

Edel, Matthew and Jerome Rothenberg, *Readings in Urban Economics*, London: Macmillan, 1972.

Eide, A., R. Jenison, L. Mashaw and L. Northup, *Engineering: Fundamentals and problem-solving*, New York: McGraw-Hill, 2002.

Environmental Protection Agency, Idle Reduction, Washington, D.C.: EPA, 2008.

Geddes, Spencer, *Estimating for Building and Civil Engineering Works*, 9th Edition edited by John Williams, London: Butterworth Heinemann, 1996.

Heggie, Ian G., *Transport Engineering Economics*, London: McGraw-Hill, 1972.

Hicks, Tyler G., *Standard Handbook of Engineering Calculations*, London: McGraw-Hill, 1972.

Institute of Civil Engineers, *Report of Joint Committee on Location of Underground Services*, London: ICE, 1963.

Little, W. et al (eds.), *The shorter Oxford English Dictionary on Historical Principles*, vol. II, Third Edition, Oxford: Clarendon Press, 1933.

Louis Berger International, Inc., *Final Report on the Study of the Transportation Plan of Cameroon. Vol. 3*, East Orange, N.J.: LBI, June 1986.

Marlow, M. and G. Maycock, The effects of zebra crossings on junction entry capacities, Crowthorne, Berkshire: Transport and Road Research Laboratory, 1982.

Ministry of Planning, Republic of Cameroon, *Fourth Five-year Socio-Cultural and Economic Development Plan 1981-1985*, Yaoundé: Cameroon News and Publishing Corporation, 1980.

Ministry of Planning, Republic of Cameroon, *Fourth Five-year Socio-Cultural and Economic Development Plan 1986–1991*, Yaoundé: Cameroon News and Publishing Corporation, 1985.

Ministry of Public Works, Republic of Cameroon, *Road Traffic Census Report, 2010-2011*, Yaoundé: Cameroon News and Publishing Corporation, 2012.

Ministry of Transport, Memorandum no. 780. Design of Roads in Rural Areas, London: HMSO, 1961.

Ministry of Transport, *Urban Traffic Engineering Techniques*, London: HMSO, 1965.

Ministry of Transport, Republic of Cameroon, *Transport Statistics Bulletin*, Yaoundé: Cameroon News and Publishing Corporation, 2012.

National Institute of Statistics, Transport statistics, Transport Statistics Bulletin, Yaoundé: NIS, 2012.

Nuland, Sherwin B. *How we Die: Reflections on Life's Final Chapter*, London: Chatto & Windus Ltd., 1994 (ISBN 0 7011 62775).

OUP, *Concise Oxford English Dictionary*, 9th edition, Oxford: OUP, 1995.

Republic of Cameroon, Declaration of the Government Urban Strategy, Yaoundé: Government of Cameroon, January 2000.

Shutung Mungwa, M., U. Chinje Melo and Paul Tamajong, *Report of a Technical Inquiry on a Collapsed Building in Yaoundé*, commissioned and financed by the Cameroon Society of Engineers April 1998, unpublished.

Tschebotarioff, Gregory P., *Soil Mechanics, Foundations and Earth Structures*, International Student Edition, London: McGraw-Hill, 1951.

Webster, F.V., Transport in Town, *Journal of Transport Economics and Policy*, May 1986, p. 129.

Webster, F.V. and R.F. Newby, Research into Relative Merits of Roundabouts and Traffic-signal Controlled Intersections. *Proceedings of the Institute of Civil Engineers*, London: ICE, January, 1964.

Wikipedia, *Costa Concordia* disaster, retrieved General Motors Corporation, retrieved 5 April 2012.

Wikipedia, General Motors Corporation, retrieved 5 April 2012.

Woods, K.B., D.S. Berry and W.H. Goetz (eds.), *Highway Engineering Handbook*, London: McGraw-Hill, 1960.

World Road Congress, *Report of the Proceedings of the XVth World Road Congress*, Vienna: World Road Congress, 1979.

Annex

Because engineering is not dogmatic, we have decided to provide real evidence in pictures about what engineers have done in recent years to rescue the situation of streets and roads. Examples are from Africa, Europe and South America.

The unexpected collapse of the bridge over the Mungo in Cameroon on 1 July 2004 left passengers and goods stranded on both sides of the river. Using this bridge, drivers journey about one hour (112 km) to the nearest major city of Limbe. Without the bridge, the only alternative road to the same city would cover 299 km with a journey time of about five hours. Engineers proposed a temporary bridge using a large pontoon from the Cameroon Engineering Shipyard Company. Within a few weeks, the temporary bridge was up and running (Photos A, B).

The unpredictable nature of mines came to the forefront in Copiapo, Chili, in 2010 when 33 miners were trapped underground at a depth of 700 metres (2,300 ft). The world held its breath and prayed. Engineers mounted a global rescue effort that paid off. On 13 October 2010, all 33 men in the Chilean mine were brought to the surface using a winching operation that had been designed by engineers (Wikipedia retrieved 22 October 2013) (Photo C).

In 2013, the *Costa Concordia,* an Italian cruise ship carrying 4,252 passengers from all over the world ran adrift on the first leg of a cruise around the Mediterranean Sea. Thanks to engineers and naval architects, the parbuckle salvage operation of the ship was successfully carried out on 16 September 2013. The wrecked ship with a displacement of 50,000 tonnes was set upright in the early hours of 17 September 2013. It is on record as one of the most extraordinary engineering achievements of all time (Photo D).

The other diagrams depict the imaginary building line which restricts the distance between the centre line of the street and buildings. Some street pavement are destroyed by contractors (such as in Tsinga, Yaoundé, and other locations) when laying water pipes, communication cables or other underground service lines. Often, pipes and cables are placed in disorder and not documented for purposes of maintenance or future rehabilitation of streets (Diagrams A-E).

The City Street in Camaroon

Photo A. Installing a temporary pontoon bridge.

Photo B. The collapsed steel thrust bridge.

Costa Concordia disaster

From Wikipedia, the free encyclopedia
Jump to: navigation, search

Costa Concordia disaster

Costa Concordia during salvage operation, July 2013

2010 Copiapó mining accident

Photo D. Rescue efforts at San José Mine near Copiapó, Chile, 10 August 2010.

The City Street in Camaroon

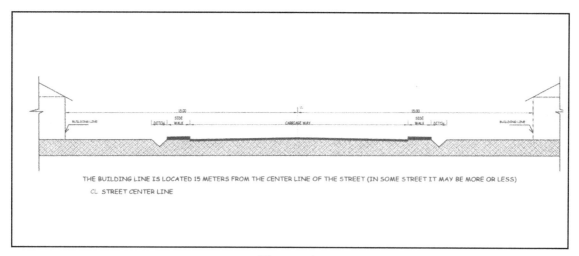

Diagram A.

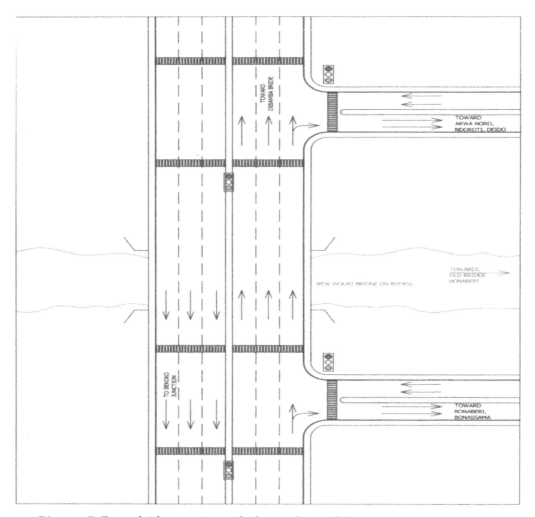

Diagram B. Future bridge crossing on the bypass from Bekoko Junction to Dibamba Bridge

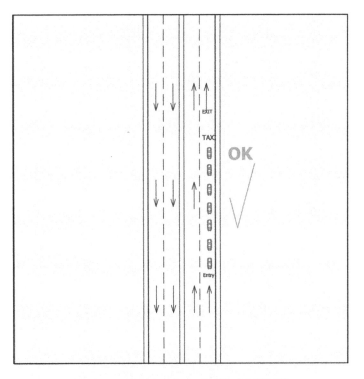

TAXI PARK SINGLE QUEUE
NO ENTRY OR EXIT FROM THE SIDE

Diagram C.

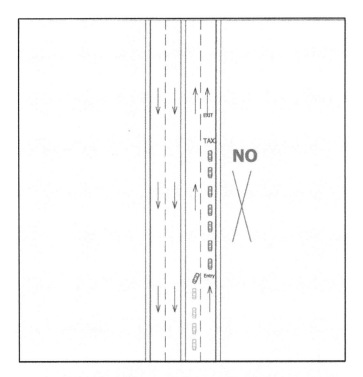

TAXI PARK·WRONG QUEUING
CREATES CONGESTION

Diagram D.

The City Street in Camaroon

Photo E. Street destroyed during maintenance of water pipes.

Photo F. Tsinga Street, Yaoundé, built to last but destroyed during laying of water pipes.

Proposal for traffic laws and ordinances

In some countries, states have the main responsibility for providing the legal basis for traffic control through the State Highway Code. These codes generally provide the following:

(a) General rules on driving and walking applicable to all public highways (speed limits, rights-of-way, making of turns, meaning of signals and signs, etc.);

(b) Rules on accident reporting, vehicle equipment, licensing of drivers, financial responsibility, vehicle registrations;

(c) Enabling legislation to permit local authorities to adopt regulations applicable to specific locations (speed zones, one-way streets, parking control, turn prohibitions); for example, in London, the Royal Borough of Kensington and Chelsea published standard conditions for street trading.

The state also has the responsibility for providing adequate and enforceable legislation on the following:

(a) Housing construction in cities and towns;

(b) Definition of the building line within and outside of cities;

(c) Norms for the occupation of city streets;

(d) Management of city streets;

(e) Management of solid and liquid wastes;

(f) Implementation of regulations and enforcement of laws (especially for traffic violations and littering);

(g) Road safety;

(h) Management that ensures the removal of debris and broken-down vehicles from highways and city streets.

Proposal for neutral street names

The table below includes neutral street names that can be used to avoid politicizing such names.

1	ACACIA	33	FICUS
2	AGRAVE	34	FIRMA
3	ALNUS	35	GONA
4	ABO	36	GARDEN
5	ALOE	37	GRAPE
6	ARBUTUS	38	GANG
7	AROA	39	GUAVA
8	ARENGA	40	HOMA
9	AUCUBA	41	HOKO
10	AKO	42	IROKO
11	ATALA	43	IBANA
12	AVO	44	KOLA
13	BATANE	45	LAWA
14	BEGEH	46	LOLO
15	BOMBAX	47	MANGO
16	BINGO	48	MAMA
17	BAMBU	49	MBIA
18	CANNA	50	ORANGE
19	CLUSIA	51	PRINCE
20	CARICA	52	POWER
21	COFFEE	53	PIONEER
22	COCOA	54	PIN
23	CARBON	55	PINEAPPLE
24	CAMWOOD	56	PIPE
25	DOGA	57	PIRATE
26	DAM	58	PALM
27	DONGO	59	PARK
28	DOGMA	60	QUEEN
29	EBO	61	ROOTS
30	EBOSSO	62	ROMA
31	EBAM	63	ROAD
32	FALA	64	RAM

Diagram E

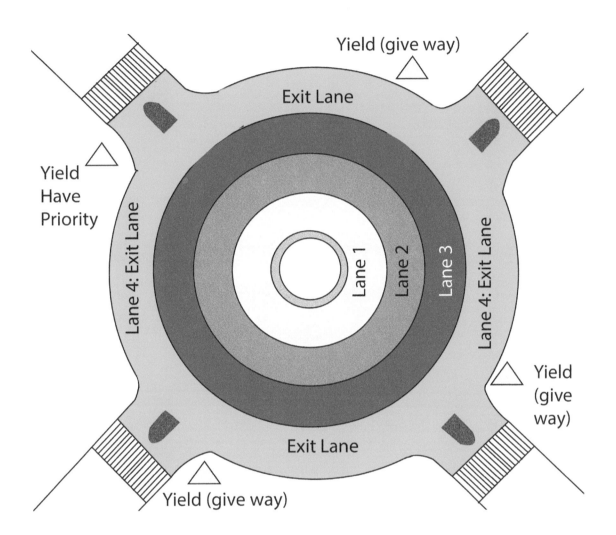

CPSIA information can be obtained
at www.ICGtesting.com
Printed in the USA
BVHW022104230421
605722BV00012B/1769